# THE MATERIAL ADVANTAGE

## BERNARD J. BULKIN

# THE
# MATERIAL
# ADVANTAGE

A Techno-Economic Look at the Past
and Future of Competitive Advantage
of Nations Through Materials

# BERNARD J. BULKIN

Author of *Crash Course* and *Solving Chemistry*

First published in 2024 by
Bernard J. Bulkin, in partnership with Whitefox Publishing Ltd

www.wearewhitefox.com

ISBN 9781916797840
Also available as an eBook
ISBN 9781916797857

Designed and typeset by Koki Design
Cover design by Dan Mogford
Project management by Whitefox

**TO VIVIEN**

*A quarter century of love, companionship,*
*and critical reading of my work*

# TABLE OF CONTENTS

*Galileo was not only the discoverer of the law of falling bodies...
but also an inventor; besides, he was not the only one in his time
who was seized by the new spirit. On the contrary, historical
accounts show that the matter-of-factness that inspired him
raged and spread like an infection. However disconcerting it
may sound nowadays to speak of someone as inspired by matter-
of-factness, believing as we do that we have far too much of it,
in Galileo's day the awakening from metaphysics to the hard
observation of reality must have been...a veritable orgy and
conflagration of matter-of-factness! But should one ask what
mankind was thinking of when it made this change, the answer
is that it did no more than what every sensible child does after
trying to walk too soon; it sat down on the ground, contacting
the earth with a most dependable if not very noble part of its
anatomy, in short, that part on which one sits. The amazing
thing is that the earth showed itself to be uncommonly receptive,
and ever since that moment of contact has allowed men to entice
inventions, conveniences, and discoveries out of it in quantities
bordering on the miraculous.*
Robert Musil, *The Man Without Qualities*

# PREFACE

Why are there rich countries and poor countries, and why have vast differences in their wealth persisted over millennia and into our own times? Moreover, how do some countries break out of a position of competitive disadvantage and become wealthy, while others remain poor, or occasionally shift from being a global power to becoming a minor, not particularly wealthy country? These sorts of questions have concerned many economists and economic historians. They have come to the fore again in the late 20th and early 21st century in the context of what has been touted as the era of a globalised economy. The outcome of this globalisation? Continued inequality and more intense national competitiveness than we have seen for some decades.

The national enterprise – an assembly of research, invention, engineering, capital, individuals and governments willing to take risk – sometimes achieves results beyond those of their local or global competitors. For this to happen a lot of factors need to be in place. After all, there are clever, ambitious people everywhere, in every country, yet some systemic factors favour enterprise, while others discourage it.

This book attempts to examine this question of enterprise generating national competitive advantage through one lens, the production of manufactured materials – paper, steel, cement, plastics, electrical cables, transistors, clothing, food packaging and many more. The thesis I am going to put forward is that the ability to transform raw materials – trees, ores, crude oil – into manufactured materials has been a source of competitive advantage of nations for thousands of years, still offers such an advantage, and will continue to do so into the future. Indeed, it is a far more important source of competitive advantage than having abundant raw materials.

Materials as I use the term in this book are the stuff for buildings and infrastructure – for example, cement, steel, glass; for food packaging – aluminium, plastic, cardboard; for conveying and saving information – paper, film, magnetic and electronic media, optical fibre; for providing, transmitting and storing electricity – copper, silicon; for trains, planes and vehicles – steel, aluminium; for the clothes we wear – linen, wool, Nylon. Materials are pervasive in modern life but have been present and important for millennia.

Now, if you are wondering whether that is something you want to read about, let me assure you that it is not a dry tale of economic principles and graphs, though there are certainly lots of principles in here and a few graphs. Economic advantage from materials is full of good stories: how canned food was invented long before the can opener (early cans had markings showing where to place chisel and strike with hammer); by contrast, Elisha Otis invented the elevator brake decades before there was a building tall enough to need one; how a patent agent asked to help a French inventor gain coverage in England stole the invention for himself; Muslim warriors threatening to kill prisoners in Central Asia unless they could teach them something useful, leading to widespread use of paper in the Islamic world; the battles between the wool and cotton industries in England, being clashes between powerful political and social forces; why it was forbidden by law at one point to be buried in a linen shroud; why colonial powers built railways in Africa and Asia but not electricity grids, the consequences of this enduring to the present day.

There is plenty of science and engineering as well as invention as part of this story, but I have kept the detailed science out of the main text and you will find it in a series of 'tech talks' that you can read or not read, depending on whether you really want to know more about how aluminium metal is made from bauxite, the chemistry of concrete, the inventions of the Industrial Revolution, or which material I consider to be the most amazing development of the 20th century, etc.

The origins of this book are in a series of conversations I had with Dr Paul Rutter. He approached me with the idea of our jointly writing a book about the history of materials, and from our discussions an idea formed in my head about materials and competitive advantage. Of course, that was a different book from the one he wanted to write, so I was on my own. The seed of the idea, however, goes back many years earlier, when as part of a small group of executives I had the pleasure of listening to a talk by Professor David Landes of Harvard University, which in turn inspired me to read his classic work *The Wealth and Poverty of Nations*. That led me to many other views of national competitiveness, from Michael Porter, Eric Hobsbawm, Paul Krugman, Niall Ferguson, Jared Diamond – leading historical and economic thinkers. None of them seemed to fasten on the importance of materials, nor did the excellent books about materials, such as those by Mark Miodownik, Stephen Sass and Vaclav Smil talk about competitive advantage in any explicit way. As I wrote the book, the centrality of enterprise came to the fore, both in the national conditions that enable it and in the amazing individuals who, over centuries, made something extraordinary happen.

I am very grateful to several people who read early drafts and gave me a large number of very helpful comments, including Paul Rutter, Chris Anastasi, Mikko Arevuo, Charles Frank, and, over the course of many conversations, Professor James Utterback. The book is quite different from the draft they read because of their useful suggestions. When I started preparing the Tech Talks, I had help from Victoria Stomberg, and she also worked on timelines with me. The Tech Talks on aluminium and cement are based on her research and writing.

Bernard J. Bulkin
London
July 2024

# 1. MATERIALS AS A SOURCE OF ADVANTAGE
## (AND AN INTRODUCTION TO THE PLAN OF THIS BOOK)

There was a time when people in most temperate climates could be clothed, albeit sometimes minimally, by killing animals and using their skins as coats and shoes. But the number of animals available for this use, even with domestication, limited the population that could survive in harsh climates. While linen from flax had been woven in ancient Egypt, and probably for several millennia before that, for use as shrouds and decorative textiles, it was the scale achieved by advances in spinning wool that changed the size of the population that could live in colder climates. Still, woollen fabrics are costly in terms of human effort. Then the massive growth of cotton as a crop, for example in the southern US, that could be processed and made into cloth by machines, expanded the amount of clothing by orders of magnitude, still with a substantial labour requirement in both agriculture and processing. In the 20th century synthetic polymeric materials from oil and gas made another step change down in the cost of clothing and in the labour needed to produce it.

And for food. Until the early 1800s food could only be preserved by techniques such as smoking, pickling or salting, processes that work only for a very limited set of foodstuffs. The invention of 'canning' by Nicolas Appert (developed between 1800 and 1810) using glass containers, and further advances shortly thereafter using sealed metal containers ('tins'), led to fruit and vegetables being available on naval vessels,[1] and ultimately to a transformative change for the broader population – particularly in France and Britain where these advances occurred – in food over seasons and geography. While the use by sailors was crucial for success of empires, it was not just a military advance. For societies with long winters during which no food could be grown, canning meant the availability of much greater varieties of food, and hence essential vitamins throughout the year (even before the role of vitamins in health was understood).

And advances in information materials. Even when the Chinese could only produce 200 sheets of paper a year, there was still a growth in literacy that increased the recording of information and ideas. Then paper from rags led to a step change in the quantities available, and subsequently wood pulp as a feedstock for the paper industry became mass produced. For example, every person in Sweden learned to read the Bible and used this reading skill to push forward an early industrial society because they could follow written directions. When newsprint became so cheap that a daily newspaper cost just pennies, the spread of information and opinions profoundly affected how people demanded that their societies were governed. And when film added pictures of events, of political leaders, of war and of the end of war, societies with a free and vibrant press changed the way they were governed, to the disadvantage of those without such media.

Each of these is an example of how progressive innovation in materials – fabrics, tin-coated steel containers, paper, film – changed the life of the populations that had these materials. When in addition nations could export their manufactured materials to increase wealth, there was positive feedback into the economy that made the country more powerful still.

Societies at the forefront of changes in development and exploitation of materials have advantages over others, and they can use this advantage in many different ways.

The thesis of this book is that repeated advances in materials for crucial human activities have driven society and particular economies forward over millennia, often at the expense of others. While other books[2, 3] have described in detail how some of these advances have occurred, for example the production of steel, cement or electronic materials, or described some of the history and impacts of certain materials in our lives,[4] they have not focused on how successive advances in such areas as food packaging, building materials or grid electricity have impacted human civilisation. In particular they have not explored how these advances have given the inventors/developers/exploiters advantage over others. Michael Porter,[5] in many ways building on the much earlier foundations laid by David Ricardo, has shown how political systems can lead to substantial competitive advantage for nations, but his book, focusing on the 20th century, heavily discounts the possibility that certain technologies, or material processes, can be used to the benefit of one country.

It is not sufficient to just assert that there is potential competitive advantage; rather it is necessary to set out specifically what sort of advantage can accrue from the ability to use novel materials, then demonstrate each sort with specific examples. That is what this book attempts to do. Materials offer the possibility of competitive advantage, but it is how a society actualises that potential that makes the advantage tangible, and in some cases sustainable in the longer term.

## THE ROUTES TO COMPETITIVE ADVANTAGE THROUGH MATERIALS

Here is a set of propositions. Novel materials, when combined with design and manufacturing, may enable competitive advantage through at least six possible routes:

1. Strengthening the ability of a society to prosper internally through better access to cheap materials, or materials that allow for the growth of industry or the welfare of an increasing population.

2. Producing something valuable far in excess of internal (home market) needs, generating a source of wealth flowing into a country, wealth that can then be deployed to further develop new sources of advantage or improve living conditions for the home population.

3. Employing materials for superior armaments or technologies to support military operations, leading to conquest or deterrence.

4. Using the advantage of material availability to produce public goods which in turn convey long-term advantage.

5. Enabling, by scientific research and engineering, the transformation of basic materials into new materials with completely different uses, so that a country with a substantial investment in scientific research and engineering excellence inevitably moves ahead of others, as does a society that encourages and rewards invention as well as entrepreneurship.

6. Living according to principles of sustainable development, i.e., living within environmental limits, with reduced inequality through the availability of necessities at low cost, and with a well-informed populace so as to have a strong capability for good governance.

Of these, routes 1 to 3 are fairly obvious, and many examples of them will be discussed. When the British produced high-quality wool for clothing and other textile uses, it allowed citizens to live more comfortably in harsh weather. As this production exceeded internal demand, exports became a source of wealth that could be used to finance empire and the Industrial Revolution. It has already been mentioned that canning supported the British and French navies, and there are numerous examples of materials for armaments, from weapons to armoured vehicles. Steel, concrete and glass, used in construction of skyscrapers, allowed Chicago and New York to become dominant US cities in the early 20th century, leaving other competitor cities such as Philadelphia, Cleveland and Pittsburgh behind. Aluminium and its alloys played roles in commercial and military aircraft, in electrical transmission cable and in many household applications.

Route 4 is characteristic of a society where the government understands the role of certain public investment decisions in broadening prosperity and strengthening national unity. Economists use two qualities when defining a public good: it must be non-rivalrous and non-excludable. The former means that if one person benefits from the good, it does not reduce the amount available to others, and the latter means that once the good has been provided, no one can be stopped from benefitting from it. Public goods often depend on materials for their realisation, and they provide a country with a substantial source of competitive advantage. This will recur several times in the discussions that follow. Perhaps the earliest case which is familiar to many is Roman roads, used to facilitate trade as well as military movement. More recent examples would be the use of aluminium alloys to transmit grid electricity, and the use of asphalt or concrete for the construction of the interstate highway system in the US as well as the British and German motorway systems. The major national railway systems were also conceived of as public goods. Flood defences are another example that is heavily dependent on materials. These are materials-based infrastructure, available to all (even if sometimes at a modest fee such that the cost does not exclude a significant number of users), the use of which does not diminish their availability to others.

Route 5 has both ancient roots and modern implementation at scale. While the Chinese were really doing technology rather than science long ago, they did invest heavily in the creation of new materials and early ideas for manufacture of these materials. An example is the early development of paper. Like materials for clothing, paper itself underwent several successive technological revolutions, each leading to greater availability at lower cost. The 20th century saw the emergence of synthetic polymers, primarily from academic and industrial laboratories in a few countries with substantial scientific establishments. These offered revolutionary new materials for information storage, clothing, containers, etc. The industrial laboratories became especially skilled at inventing new products and manufacturing processes for the material world.

There are a series of great advances that have come through use of investment in science to repurpose known materials to new uses, for example, copper from tools and vessels to wires, silica from glass containers to quartz crystals to fibre optics to silicon chips, iron to steel and the development of many different steel alloys suited for different applications, as well as the evolution in magnetic materials, starting with iron permanent magnets, that came through understanding of the materials science of magnetism. Paper companies invested heavily in research to develop a range of products beyond the use of paper for recording information, including kitchen towel, toilet paper, tissues and menstrual products. Nylon started as a polymer for hosiery and quickly evolved to hundreds of diverse civilian and military uses. Other polymers such as the film material sold as Mylar are likewise used in very diverse fields.

Invention of new manufactured materials, development of the processes to make these in massive quantities, and understanding how the fundamental properties of these materials can lead to new uses has happened in a select subset of nations over the last two millennia. A world-class scientific and engineering establishment is crucial for obtaining broad-based competitive advantage through materials, whether for civilian or military applications. Research can also be effective at reducing or eliminating the advantage of a competitor society. It is not surprising that the major powers in World War II were Germany, the United States, Russia and the United Kingdom. These countries, along with France, were the leading scientific centres in the world in the first half of the 20th century. As one measure, between these five countries they won more than three quarters of the Nobel Prizes in Chemistry and Physics from the start of the Nobel Prizes at the beginning of the 20th century until the end of World War II.

Route 6 is the most underappreciated, and until the latter part of the 20th century would probably not have been mentioned in the context of this discussion. Even as ideas of sustainable development emerged, many equated it with environmental protection, or even more narrowly with managing climate change. It is now recognised that this is just a part (albeit an important one) of what sustainable development means. What is the more expansive meaning? Considering the array of choices for government, and more broadly for our society, sustainable development is a central organising principle that can be applied to make these choices. It says that integrating economic, social and environmental concerns over time, not crude trade-offs but searching for mutually reinforcing benefits, is the way to approach problems. Sustainable development promotes good governance, healthy living, innovation, lifelong learning and all forms of economic growth. Sustainable development tries to build social harmony and seeks to secure the prospects for everyone to lead a fulfilling life. This means that societies living according to the principles of sustainable development strive to reduce inequality and increase intergenerational opportunities.

In this regard, decisions must value nature, ensure that the polluter pays, that governance is participative, and that where there is doubt, particularly as it stretches across generations, society adopts a precautionary approach. When it comes to markets, sustainable development tries to work for efficient economies through transparent, properly regulated markets that promote social equity alongside personal prosperity.

Looking at sustainable development in this broad context, one can see places where materials developments might offer significant competitive advantage to a society that makes its choices using these principles. There was a time when small populations, having exhausted the raw materials needed to survive, picked up and moved elsewhere rather than innovate or become more efficient. Later, as populations began to urbanise the demands on materials changed. Economists in the 18th century were divided as to whether the future was more optimistic (Adam Smith, who thought that there would be a higher quality of life for more people) or pessimistic (Thomas Malthus, who believed that as soon as there was a little more wealth birth rates would go up to compensate). As will be seen in subsequent chapters, technological innovation ultimately defeated Malthusian ideas.

## MATERIALS IN SOCIETY

Societies often start by coping with the demands of an increased population – be they for mobility, shelter, clothing or food – by trying to use materials that emphasise quantity and low immediate cost without worrying too much about waste or environmental damage. The last several centuries, however, have seen repeated cases where materials innovation allowed a bridge to a more sustainable solution. Using these innovations has not been optional. As advances in medicine, hygiene, food quality/quantity and safety have removed limits to population growth, it has only been possible to feed and clothe ourselves through innovation and improved productivity. By reducing such constraints as infant mortality and water-borne diseases, the 'materials system' is driven to respond. The ways that it does this are described later in this chapter.

This will be illustrated with several examples, in energy, agriculture and transport. It can also be seen in the use of innovative materials to make information broadly available in our society. The materials that enabled cheap newspapers and film for capturing and reproducing pictures completely transformed the governance of democratic societies. If there was any doubt about this, one need only look at how totalitarian societies suppressed and continue to suppress the use of these materials.

Still, while in this book the great advances in materials for packaging, clothing, food preservation and electronics, many of them based around fossil fuel derived plastics, will be described, and even celebrated, it is with a recognition that this has often been done without due regard to the sustainable development principle of living within environmental limits. The consequences may have taken some time to be recognised, but they are abundantly clear, and change needs to happen. Nylon and polyester, Bakelite and pop top cans were great inventions. The book will describe how competitive advantage was achieved through the scale-up and deployment of such innovations, indeed how they made life better for people in measurable ways. But this celebration must always be with both eyes open to the violations of the principle of living within environmental limits that occurred and continue to occur. In Chapter 9: The Future Materials Enterprise, it will be asserted that the 21st century requires massive change in both the materials we use and how we manufacture them, that there is ample opportunity to make this change, and that once again some societies will achieve competitive advantage by doing so.

Finally, there are certainly examples of materials which, even in antiquity, were fairly universal and though necessary they did not convey any competitive advantage. The formation of clays into pottery is very widespread in the archaeological record – different shapes, decoration and methods of formation – but most societies found ways to make ceramic pots. This book is not going to say anything more about such examples. In any discussion of competitive advantage, there will always be certain things which are necessary for a society to function but relatively easy to obtain, while there are others which are only available to a select group. Of those who have these resources, even fewer will know how to use one or more of the six routes to achieve a competitive advantage.

## BUT WHAT ARE 'MATERIALS'?

To take this argument forward, it is necessary, or at least desirable, to define the scope encompassed by this word 'materials'. There is not a precise or even an accepted definition of materials. Adopting a pragmatic approach, a useful way forward is to make a definition by saying what is inside the scope of materials and what is outside for the purposes of this book, and not worry too much about the borders where there is ample possibility for disagreement.

The term 'raw material' is in common use. The raw material for an iron tool or a steel girder is iron ore, the raw material for polyethylene is ethylene, and this in turn is usually produced from oil or natural gas.

If a country has raw materials such as oil or iron ore these can be a source of wealth, but they are not necessarily a source of competitive advantage through materials. For many countries raw materials have more often been a motivation for others to exploit them. In any case, as pointed out by Piketty,[6] if a country extracts 100 billion euros of crude oil from under its land it generates 100 billion euros of increased gross domestic product (GDP) but zero increase in national income, because the stock of natural capital has been reduced by the same amount.

Materials as discussed in this book are the manufactured products, even if in a preliminary form (for example, polyester powder which can be made into textiles, packaging film or bottles) which can then be put to use in society. Pure iron is a material when it is in a form that can be used for a girder, a magnet or sintered from powder to form an engine block. Likewise, uranium ore is not a material, but purified uranium used as nuclear fuel is.

Even here it is useful to draw a boundary. Ethylene is a product of oil; ammonia is a product of nitrogen (from air) and hydrogen (often obtained from gas or oil). But the scope will not include ethylene or ammonia as materials, indeed it will exclude all gases. The utility of ethylene comes as polyethylene, and ammonia as fertiliser or other polymers incorporating ammonia, so it is these products made using the gaseous building blocks that are inside the scope of what this book includes as materials. On the border, not discussed at any length in this book but consistent with how materials have been defined, are the fuels that result from refining of crude oil such as gasoline and diesel.

Silicon chips, copper wire, glass as windows or optical fibre, these are all materials. All synthetic polymers, pretty much all metals in their useful form, solid state electronics, film, cement, alloys such as steel and aluminium alloys, ceramics (a very broad category which includes pottery, tiles, bricks, porcelain) and asphalt, these are all materials as far as this book is concerned.

While not all materials scientists agree, materials as defined in this book also include most naturally occurring substances with similar uses. So wool, cotton, leather, rubber and of course paper (usually these days from wood, at various times from rice, bark or rags) are all materials once they have been processed into a useful form. These contain polymers of biological origin, that is, they are synthesised by living systems in plants or animals. An example is cellulose, present both in paper from wood and also the main constituent of cotton.

Wool, by contrast, is a quite different polymer, keratin, something more akin to a protein containing a variety of amino acids that coil up in helical chains linked by sulfur-sulfur bonds, while leather is a polymer called collagen that has been processed from skin, also containing amino acids, but wound together in a triple helix. These polymers are chemically related to the synthetic polymer nylon. Rubber, which in its natural state is an emulsified material called latex containing proteins, starches, sugars, gums and many other components, is included among materials in the definition, but a tree, (even a rubber tree), a cotton boll or a cow are not materials, any more than a barrel of crude oil. Again, having lots of sheep or trees or fields of cotton, high clay content in soil, or a rubber plantation, can be a source of wealth, but not of competitive advantage.

Also on the exclusion list for this book, though again not on everyone's, is processed food,[7] although it is often manufactured and sold in large quantities. This is partly convention, and because no society derives competitive advantage from having better chocolate or Pringles crisps, although possibly they might from baby formula or vitamin-enriched foods. However, the containers and packaging for the food are included as materials.

These days, one thinks, and correctly, that materials must be manufactured at great scale in order to have an impact on the economy of a society. Scale is not always measured in weight or volume, it must include an element of value. But to give an idea of scale, the US produces about 500 billion sheets of paper from trees every year, of which about half eventually get recycled; about 180 billion aluminium cans are sold each year containing beer or soda (with a high percentage of recycling in certain countries); 24 billion kilograms of cotton are produced for clothing each year, with a water requirement of about 10,000 litres per kilogram. To achieve this scale, great enterprises must be built. This book is about the conditions that allow this to happen and the stories of the individuals and societies that built these over the last two millennia.

Another insight into the decision of what to include as materials comes from looking at the components of the Standard and Poor's Materials Index, the companies in the S&P 500 that derive a substantial portion of their business from materials.

| COMPANY | MATERIALS MANUFACTURED |
| --- | --- |
| Albermarle | Lithium for batteries, key polymer components, fire safety materials |
| Air Products and Chemicals | Industrial gases and membranes for gas separation |
| Amcor | Packaging for food and pharmaceuticals |
| Avery Dennison | Labels for packaging, adhesives, tapes |
| Ball | Cans for food, aerosols, beverages |
| Celanese | Polymers and building blocks for polymers |
| CF Industries | Products containing nitrogen, especially fertilisers |
| Corteva | Agricultural chemicals and seeds |
| DuPont deNemours | Construction materials, polymers |
| Dow | Polymers |
| Ecolab | Food and medical safety products, refinery chemicals |
| Eastman Chemical | Numerous polymers for application across many industries |
| Freeport McMoran | Mining of copper, gold and molybdenum |
| FMC | Agriculture equipment, lithium, various polymers cosmetics, chemicals enhancing taste and smell |
| International Paper | Paper products for packaging and printing/writing |
| Linde | Industrial gases |

| | |
|---|---|
| Lyondell Basell | High volume chemicals |
| Martin Marietta Materials | Concrete, cement, asphalt, aggregates for buildings and roads |
| Monsanto | Agrochemicals and agricultural biotechnology |
| Mosaic | Mining for fertiliser products |
| Newmont Mining | Gold and copper mining |
| Nucor | Steel |
| PPG Industries | Paints and coatings |
| SealedAir | Food packaging, bubble wrap |
| Sherwin Williams | Paints and coatings |
| Vulcan Materials | Aggregates, asphalt, ready mixed concrete |
| Westrock | Paper and packaging |

The definitions discussed above would not encompass all of these, for example excluding the pure mining companies, as well as industrial gases and polymer precursors, but certainly includes the output of most of these companies.

In the 20th century materials scientists and engineers began to speak of functional materials. Now all materials that are in use in society are performing a function, whether it is protective clothing, food preservation, structural support or transmitting information in the form of the written word. The term functional materials goes beyond these sorts of basic functions to include electronic materials such as transistors, optical materials for lasers, solar cells, fibre optics, chemically active materials that might have both a structural and catalytic function, for example the catalytic converters in automobiles, and biological or biomedical materials, again having an active role in mediating a biological system not just performing a structural function such as a dental repair or hip replacement. A high proportion of the frontier research in materials science and engineering is about such functional materials.

## MATERIALS AS COMPONENTS OF TECHNOLOGICAL SYSTEMS – AND THE IMPORTANCE OF SYSTEMS THINKING

Competitive advantage – for a country, a city or a multinational company – is a possible outcome of a system in which the use of a manufactured material is embedded. The processes used to manufacture a product from a raw material are what is meant by technology.

What is meant by a material embedded in a system? It is necessary to introduce the idea of systems thinking, which is fundamental to the more detailed description of materials and their uses. Systems thinking is an especially useful tool for analysing what is happening in the world around us.[8] Essentially it says that when something is referred to as a system – the financial system, ecosystem, political system – it implies more than just interconnectedness. It means that the system functions as a whole, that the output of the system is more than just the sum of individual outputs, that an enhancement or a malfunction in any one place will affect the entire system. Systems thinking tries to uncover how the parts of a system interrelate, how the system sits in the context of a larger system in which it is embedded (for example the education system in the US political system or its counterpart in the Chinese political system) and how the system behaviour will evolve over time.

While books such as *The Fifth Discipline* give the basic principles, and apply these ideas to organisations, Howard and Elisabeth Odum have developed the ideas of systems thinking for systems involving energy, materials and information.[9] Systems thinking developed in the decades following the work of Jay Forrester and Peter Senge. Donella Meadows was a leader in taking these ideas and putting them in the context of sustainability.[10] Porter, in *The Competitive Advantage of Nations*,[6] has also highlighted the importance of systems thinking, though from a broader economic perspective. His overarching system includes factor conditions, related/supporting industries and demand conditions, all of which will feature in the discussion of materials. However, he places considerable emphasis on firm structure and the political/regulatory systems that enable firms to grow and compete, which will be dealt with less extensively in the discussions to follow. Much of the research on systems thinking in the period up to 2015 can be found through the discussion and references in an article by Arnold and Wade.[11]

In the systems analysis discussed here, there is the embodiment of processes for making (manufacturing) a material from its raw materials, such as a steel mill. There is often a subsequent step of research and development that leads to a variety of products, such as stainless steel or other steel alloys with particularly desirable properties. Then there is a use for that material, for example in buildings or trains or cutlery. Arrows connect these points and represent the drivers (or inhibitors) of change, for example, cost reduction, speed of supply, advertising or availability of energy.

The systems diagrams are also a representation of one of the most important observable phenomena in the role of materials in the economy, namely positive feedback. As a company or nation develops one product from a raw material and brings it to commercial scale, it sees opportunities to reduce costs. The lower costs in turn open markets for additional products based around the same material. The wealth generated from the industrial success leads to more disposable income for the middle class, in turn generating demand for new products. These sorts of positive feedbacks are at the heart of most of the examples described in this book.

## RAW MATERIALS, RESEARCH AND ENERGY AS INPUTS

Consider, at an abstract level, the system by which materials flow into and through society. In Chapters 3 to 8 there will be many specific cases of these systemic flows. Raw materials (ores, oil, etc.) go through a process of manufacturing; initially in small amounts after a process is invented in the laboratory, then at scale. The process is itself the product of research and development (R&D), whether in an academic, large industrial or small entrepreneurial enterprise, creating the technology for producing a range of products. In a famous example, the process for economically producing pure aluminium from bauxite ore was developed independently by Charles Martin Hall in the US and Paul Héroult in France. This was an exceedingly difficult research achievement that had eluded many other researchers. Then Hall raised the capital and brought together the engineering capability to turn the laboratory demonstration into a proper commercial process.

It is clear from the archaeological record that some transformations of raw materials into a useful product are easy and have widespread occurrence around early human civilisation. Examples are the aforementioned process for turning clay into pottery and, perhaps more surprisingly, plant fibres into string and woven cloth. When the 'R&D' barrier is low, competitive advantage is usually not possible.

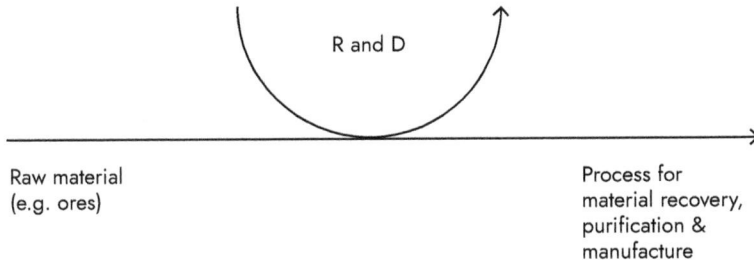

Raw material
(e.g. ores)

Process for
material recovery,
purification &
manufacture

Virtually all manufacturing processes from raw materials to finished material products require energy, and if costs are to be kept down, the cost of energy is usually a vital component. That is why the final stage of producing aluminium from its ore, in which enormous quantities of energy in the form of a reliable electricity supply are required, occurs in places such as near Niagara Falls or in Iceland, where there is no ore but there is a cheap source of electricity from hydro or geothermal sources respectively.

## BESIDES DESIRED PRODUCTS, THERE ARE WASTE AND BY-PRODUCTS OF MANUFACTURING

There can be waste products resulting from the manufacturing process, whether that is from burning a dirty fuel, by-products of the process, such as ash or slag, or simply scrap that has been cut away in the formation of a final product. Industrial leaders who are creative think of these not as waste but as products for which an economic use has not yet been found. For example, alumina and aluminium processing creates a range of waste products, the most significant being:

- bauxite residue, a sand and mud (in equal parts) slurry that contains most of the iron and silicon impurities from the bauxite along with some residual caustic soda.

- mercury emissions, which occur through refining operations as mercury naturally occurs in bauxite though the concentration varies from one batch of bauxite to another. This variability adds to the challenge of finding a common solution to reducing emissions.
- spent pot lining (SPL), the waste produced from the aluminium smelting process when the carbon and refractory lining of smelting pots reaches the end of its serviceable life.

Only in recent years have companies felt pressure to develop new uses for these waste products.

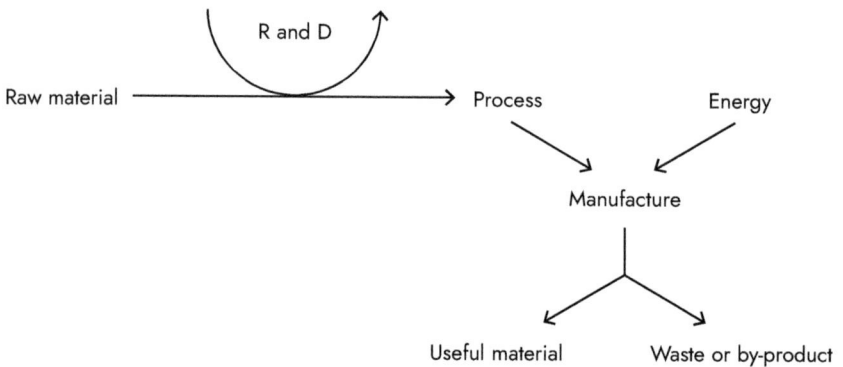

## MULTIPLE USES ARISE AND RECYCLING BECOMES IMPORTANT

The manufactured materials usually find a variety of uses. Again, aluminium is a supreme example, finding application in kitchen utensils, building materials, vehicles, aircraft, electricity transmission, packaging and machinery. Sometimes introducing a new material to fulfil an existing use is easy – the customer can see better performance, lower cost or a durability advantage. But in other cases, the customer must be helped and supported in implementation. The presence of a technical support function is often a crucial determinant in replacement of one material by another, or in a novel material performing a function that did not exist before. A company with a superior technical support offering will succeed against competitors, but technical support can also be a determinant of competitive advantage to a society. The DuPont Corporation never planned to manufacture and sell stockings from the Nylon it manufactured, but it had a fully functional mini-factory to produce these so that it could convince hosiery manufacturers, the potential major customers, of the attractiveness of its product, learn what pitfalls might occur, and solve problems as they were encountered by these customers. A novel electrical cabling alloy may be available for anyone in the world to buy, but if only US and European customers have someone that they can call on to help with its installation, they will be the ones to implement it successfully and derive the consequent advantage. It is surprising that such advantage (or more properly put disadvantage) persists even in the 21st century world of facile global communications, but it does.

While some uses of the materials require minimal processing innovation, for example converting purified metal into foil or wire, others are more complex. Occasionally there is need for inventions that don't involve the material at all in order to enable its use. For example, steel was a crucial material for construction of skyscrapers, and it will be argued in Chapter 6 that cities embracing the skyscraper achieved competitive advantage over others that were slow to build very tall buildings. These first skyscraper buildings relied on the fact that the elevator, and in particular the Otis safety braking system, had been invented earlier. Other enabling inventions, some occurring before the material technology development, some more or less coincident, others after, include the printing press, the pencil and the can opener.

Certain applications leave the materials locked up for a very long time, for example bricks, or the concrete used by Roman engineers, a large quantity of which is still in place two millennia later. Aluminium used in buildings or car engines might not last for millennia in service but will still be there for decades. But others, for example, disposable food packaging, the daily newspaper, or a piece of aluminium foil used to wrap a sandwich, immediately become waste (although the foil in such applications can be cleaned and reused, or recycled, most often it is just discarded). This can either be disposed of to landfill, sometimes burned as fuel or recycled. One of the earliest and largest examples of recycling into a new use was the collection of rags from clothing for making paper.

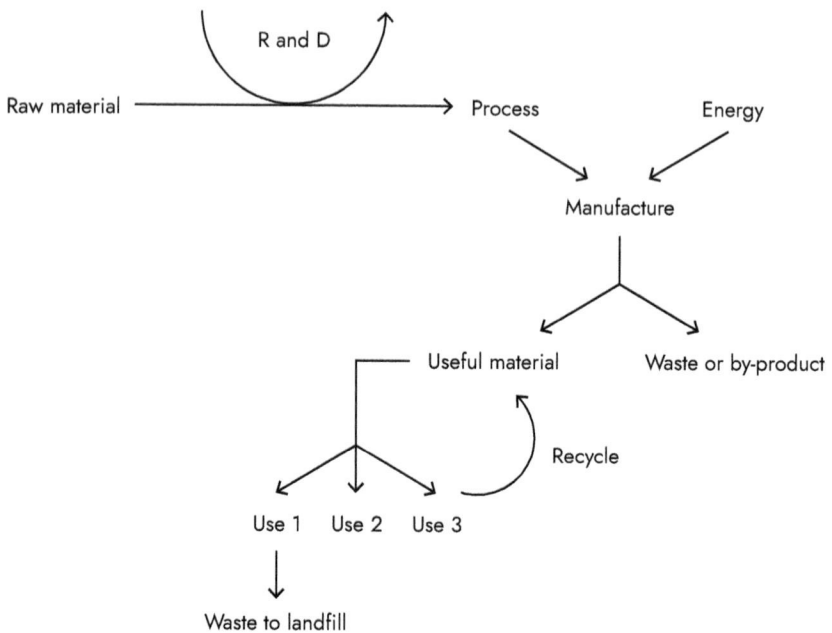

## THERE ARE DIVERSE DRIVERS OF DEMAND

The loop back from use to manufacturing can be closed by considering the drivers of demand. Sometimes these drivers are quite simple and easy to perceive, even anticipate. Other times they are more subtle or hidden. Aluminium represents one of the simple cases. The desirable properties of aluminium in terms of density, corrosion resistance, etc. were well known before the process inventions of Hall and Héroult, but the costs were too high. So the driver of demand was simply cost reduction, which in turn opened up even more uses, which had the virtuous cycle (positive feedback) of increasing scale leading to further cost reductions and more new applications. A similar situation was true for solar panels in the first two decades of the 21st century.

Another demand driver is military needs, and in this is included the need for materials having certain extreme properties as required by space exploration, such as ceramic tiles and the epoxy resins that were used to attach them to spacecraft. Already mentioned are materials for providing food or footwear to armies and navies. Armour is also a materials problem, from the metal suits of armour of antiquity to polymers such as Kevlar for bulletproof vests.

Demand for consumer goods also interplays with medical advances. About 120 years ago in Europe and the US, chlorination of water began to be widespread, and this was a major factor that led to a doubling of life expectancy in a very few decades. When the population of older people increases so dramatically, the need for materials for housing, inexpensive clothing and elderly health care likewise increases. Materials for joint replacements have become a big business in many countries, allowing older people to continue to lead pain-free lives. Given the increase in longevity in the population continuing into this century, it has been speculated that materials for adult diapers are a big growth area!

Sometimes, particularly with consumer goods, people don't know they need something until advertising tells them that they do. Advertising is thus needed by every visionary innovator, whether for materials (Nylon stockings) or devices (iPhones, ATMs). Advertising can be ethical, in that it informs consumers of the properties of materials in products they did not know existed, but it can be used to create jealousy of others or to denigrate competing materials.

Even religion can play a role in demand for materials, as it did in the desire of the Reformed church for mass production of Bibles, thus creating demand for cheap paper in Sweden and New England far in excess of what an agrarian society would need.

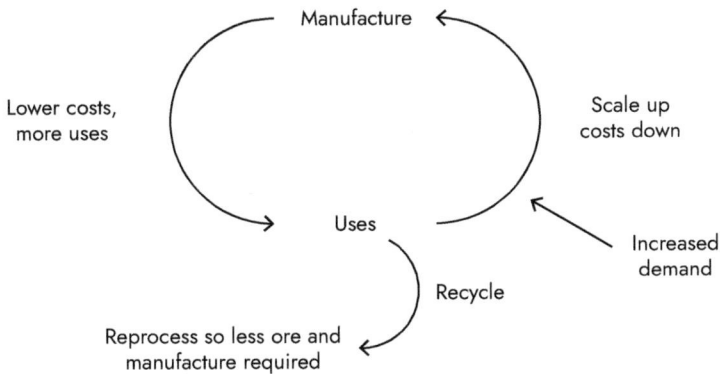

## WHAT'S MISSING?

These systems diagrams are good for setting up the story of how materials can evolve from ores or plants or animals to become essential to human society. But the diagrams do not even begin to tell the whole story. What is missing from the journey from idea to enterprise? Many things, including the context in which the development takes place – is there rule of law, is risk-taking encouraged, and other factors that will be explored in detail in the next chapter.

Also missing is the individual. Sometimes that is the scientist/entrepreneur, like Charles Martin Hall, who, having discovered in his home laboratory a process for making a small amount of pure aluminium from bauxite, went on to found the company Alcoa that dominated production of this material for half a century; or the inventor/entrepreneur Thomas Edison; or the visionary Lewis Miller, who saw the opportunity to create in Akron, Ohio, a cluster for automotive tyre development and manufacture that was crucial to the rapid growth of automobiles in America.

And capital too is a factor. Going from laboratory to manufacture at scale, such as DuPont did with Nylon, requires a lot of money. That means there are investors, whether in private or public companies, who are willing to risk their money to back an idea, knowing that many will fail to grow into great enterprises.

## A THEMATIC APPROACH TO THE MATERIAL WORLD

This book is not a history of materials. Such a history would pick some that are interesting, like paper, steel, glass or polyester, and follow them from invention to use(s). There are several books that do just that for specific materials. Rather this book is interested in a particular aspect of materials, i.e. how they give advantage to a society compared to its local or global neighbours. Earlier, six potential routes by which such advantage can occur have been enumerated. Further, the idea of materials and their uses being part of a system has been introduced, involving everything from scientific research, sources of energy, engineering process development, more research to develop new uses for the materials, dealing with waste products, and including such system drivers as medical advances, government policy, advertising and religion. Thus, materials processing and use in society have been embedded in a system, a necessary prerequisite for considering the development and maintenance of economic advantage. As the story of competitive advantage through materials unfolds in this book, these systems ideas will be implicit everywhere, and explicit in several cases.

To explore the thesis of enterprises that give national competitive advantage, materials in society are examined by six broad themes in which they are used, themes which are among the big economic factors in society. In taking this approach it will be the case that some materials grow to prominence, even dominance, and are then replaced by others, and later those are replaced again. Sometimes alternatives performing the same function co-exist for extended periods of time. True, this way of tackling the subject is going to involve some materials being considered in more than one place, for example, polyester as a textile as well as a food container – in the first of which it may have displaced cotton or linen, and the second glass or cardboard. But that is how the technological economy works.

The choice of these themes is best described as 'not completely arbitrary'. One can think of several other choices that might have been made. They are a bit self-serving of the thesis, that is, by looking into them and seeing how the six routes to competitive advantage have played out in these themes the point of competitive advantage through materials is developed. It is for the reader to decide whether it is proven. As already indicated, they are all big in terms of their economic role in society. Most of these themes go back millennia, and quite a bit is known about their history for at least the last 2,000 years, in some cases much longer. Two of them, electricity/electronics and mechanised transport, are less than 150 years old (although electricity can be considered as a particular case of energy transformation that humans have done in various ways) but their impact has been so great that it can't be ignored. All the themes exhibit an increase in the rate of change that has occurred in the last 100-plus years.

So, what are the activities that human civilisations have carried out (with the exception of electricity) since antiquity?

## CLOTHING

Our species and its close relatives spread out from their initial home in a lush hot climate in Africa to almost all parts of the planet, many of them with long harsh winters. As that migration happened, there was a greater need for clothing to protect bodies from the weather and from injury. For reasons that are not completely clear, even when people have no need of clothing for warmth or protection, they wear it for reasons of modesty. Certainly, clothing is used to attract mates (since people do not have brightly coloured feathers) and to assert or demonstrate position in society. That's fashion!

## CONTAINING AND PRESERVING FOOD

We eat things that we grow ourselves and things that we kill. Over millennia, agriculture developed and animals were domesticated for work, food, clothing and transport. But crops are seasonal, and food needs to be preserved from one season to another. People learned that crops fail periodically, so finding ways to store and preserve food from one year to the next means that the population does not starve during these periodic failures.[12] Likewise, it is necessary to contain liquids, especially water, both to be able to transport them from where they are found to where we live and to survive periods of drought. To meet our needs for living or making war far from the source of our food, food preservation in durable containers is essential.

## BUILDINGS AND INFRASTRUCTURE

Settlements are built, and these grow over time from small villages to large cities. Densification of population brings advantages in energy and land use, but it also brings challenges. In cities people need to learn how to build larger and larger structures, both for residences and for commerce. These structures must be safe to build and live/work in. They also need to be attractive to their occupants, often demonstrating by their appearance the class or wealth of those occupants. This is a distinctly human need. To bring the materials of construction, and subsequently to live and work in cities, there needs to be an infrastructure of bridges, roads, tunnels, water and sewerage.

## MOBILITY

We travel. It is built into our evolutionary heritage to explore, to trade, and just to visit. So means are required to make travel more rapid and more comfortable, as well as cheaper. Travel is also a vital component of military activity. Every form of transport humans have devised has involved materials science and engineering, but the demands of mechanised transport on materials have been both more numerous and more stringent than what preceded it.

## INFORMATION

We communicate and we want to be informed, we educate and we want to learn. So we develop not just written and spoken language but the means to make it available widely and to preserve it as an archive of our activity and creativity. Literacy has emerged as a critical component of a successful society; in turn it makes demands on materials to provide books, newspapers and, eventually, electronic media cheaply and for everyone.

## ELECTRICITY

Humans were driven to extend the day through lighting, to provide temperatures in our dwellings that were neither too hot nor too cold, and to produce devices for communication, clothes washing, food preparation, music – indeed for all things that are accepted as daily needs once more basic needs are satisfied. All these involved energy conversion processes. Over the last century it has emerged that electricity is by far the superior means to convert, transmit and store energy. There is a consequential drive to make it available cheaply and reliably to all, though this is not yet even close to universal on our planet, leaving more than a billion people at a major competitive disadvantage.

## AND WHAT MAKES US DISTINCTIVELY HUMAN

We fight, to conquer and subdue our neighbours, either to enslave them so that they labour for us, so that we can use their land, or just for the glory of it. Hence there is a demand for more effective materials of war, and for harnessing some of our technological efforts towards its effective prosecution. In certain societies a disproportionate amount of research and technology effort is diverted towards military applications. A 'domesticated' version of this fighting is sport, which is also increasingly dependent on advanced materials.

We are communal. The greatest transition in human history is associated with the growth of settlements, leading to farming, family life, housing, collective religious worship, trading with other communities, systems of law and punishment of transgression.[13]

We try to understand the world in which we live, and the worlds in which we do not live, so enterprises for scientific research and mathematics are developed, some of which have practical outcomes, and some of which are purely for our comprehension of the laws governing our universe.

We make things, transforming one thing into another. From ancient times we found ways to make pottery, needles and thread, books/scrolls, clothing. As well as making things, we decorate them, try to make them beautiful as well as functional. We use materials to produce art for its own sake, to leave a record saying we were here, perhaps so that others can find the pleasure and the intellectual challenge of our art.

We constantly strive to live longer, because we sense the finality of death – as far as we know uniquely among living things. As we succeed at this, we need more of everything, more food and clothing, more houses, more medicines, more cars. As a result, we want materials in quantity, with greater choice, provided quickly and cheaply to meet the needs of our ever (so far at least) increasing population. Perhaps we also want materials that are more durable.

The themes this book chooses to use as lenses to examine the role of materials in society are these: preserving and containing food; clothing; buildings and infrastructure; transport; information; and electricity. Using the systems approach facilitates a look at drivers of advantage involving materials such as increasing life span through medical advances, the role of a strong research and development enterprise, availability of cheap and reliable energy, the human demand to have everything better, faster and cheaper, advertising, religion and war.

After looking at this economic history and technological development of materials, it is appropriate to look forward. In a society that is interconnected by both infrastructure and politics, is it possible to sustain or even obtain competitive advantage of the sort that has occurred before? Why are there still some countries that are seriously disadvantaged, particularly with respect to certain materials? What sorts of materials advances might occur over the coming decades, and could these be a source of competitive advantage? Could creative responses to the increasingly urgent demands of sustainable development lead to societal advantage? Will the future look anything like the past?

## Notes

1. A need which had been recognised by James Lind in the mid-18th century, but which could not be met except in limited quantities because of inability to preserve high vitamin C containing fruits or juices on long voyages.

2. Stephen L. Sass, *The Substance of Civilization*, Arcade Publishing, New York, 1998

3. Vaclav Smil, *Creating the Twentieth Century*, Oxford, 2005

4. Mark Miodownik, *Stuff Matters*, Penguin, London, 2014

5. Michael E. Porter, *The Competitive Advantage of Nations* 2nd ed, Macmillan, London 1998

6. Thomas Piketty, *A Brief History of Equality*, Belknap/Harvard, Cambridge (US), 2022

7. For example, Mark Miodownik has a whole chapter on chocolate as a material in *Stuff Matters*.

8. The best introduction to systems thinking is by Peter Senge et al, *The Fifth Discipline Handbook*, Doubleday New York, 1994.

9. See, for example, Howard T. Odum and Elisabeth C. Odum, *A Prosperous Way Down*, University Press of Colorado, 2001 and the seminal work, Howard T. Odum, *Environment, Power and Society*, Wiley, New York, 1971.

10. D. H. Meadows, *Thinking in Systems: A Primer*, Chelsea Green Publishing, 2008, is an excellent guide to her work and other work on systems thinking up to her untimely death in 2001.

11. R. D. Arnold and J. P. Wade, *Procedia Computer Science*, 44, 669, (2015), available as a pdf online at https://www.sciencedirect.com/science/article/pii/S1877050915002860

12. For an early example, see Genesis 41:25–37.

13. Yuval Noah Harari, *Sapiens: A Brief History of Humankind*, Vintage, 2014, is an excellent exposition of this thesis.

# 2. NATIONAL AND REGIONAL COMPETITIVE ADVANTAGE

*Productivity isn't everything, but in the long run it is almost everything. A country's ability to improve its standard of living over time depends almost entirely on its ability to raise its output per worker.*
Paul Krugman, *The Age of Diminishing Expectations*, 1994[1]

*We are living in a dessert age. We want things to be sweet; too many of us work to live and live to be happy. Nothing wrong with that; it just does not promote high productivity. You want high productivity? Then you should live to work and get happiness as a by-product.*
David Landes, *The Wealth and Poverty of Nations*, 1998[2]

*Productivity is positively correlated with the quality of the environment.*
Philippe Aghion, Céline Antonin and Simon Bunel, *The Power of Creative Destruction*, 2021[3]

Can nations, regions or cities establish enterprises that produce a sustainable competitive advantage? History appears to indicate that this is the case, with some nations dominating large parts of the world economy for long periods of time (say, a century or longer). If the historic reality of this is accepted, then what are the conditions to make it happen? Have these conditions been more or less the same or do they change? Indeed, while economic historians seem to agree that enduring competitive advantage is (or at least, has been) possible, is there any consensus as to why it occurs?

The ideas of competitive and comparative advantage of nations as a foundation of economics go back two centuries to the great works of

Adam Smith, David Ricardo and John Stuart Mill. Ricardo laid out what he saw as a structured approach to how countries achieve comparative advantage through trade and specialisation, ideas that are still taught to students of economics. In Ricardo's approach, country A concentrates on making what it can do best (cheapest, usually) while country B does something else that it can do better. Both gain from this by focusing on where they are comparatively advantaged. The discussion in the chapters that follow is not about this important idea, but rather on a more one-sided competitive advantage. Mill extended Ricardo's ideas, particularly in his discussion of the role of world trade. The more recent formulations of these ideas discussed in this chapter build on this work while accounting for the centuries of change in society. Smith and Ricardo, living at the time of the French Revolution and the Napoleonic Wars respectively, could not have envisioned the transition in society that could occur through such things as universal free education, extension of the voting franchise, and general growth of individual freedom (at least in many countries), though Ricardo advocated for the abolition of slavery. Adam Smith certainly articulated the ideal, which he referred to as natural liberty, but it was not possible to see how to get there. Indeed, as will be seen in ideas from David Landes, Eric Hobsbawm and Niall Ferguson discussed in this chapter, in many ways this transition remains incomplete everywhere. In Chapter 1 it was asserted that wealth and advantage can come from the research and engineering that creates processes transforming raw materials into manufactured products. David Ricardo understood resources as fundamental to advantage but could not see from his vantage point in the early 19th century how much a society could create those resources beyond what they might have as a natural endowment. In this sense, the ideas of Michael Porter discussed below follow to some extent from Ricardo, and Porter describes himself as a Ricardian.

## NATIONAL IDENTITY AS A PREREQUISITE

If a nation is to become competitively advantaged, to begin with there needs to be a national identity – the citizens need to feel that they are a nation. This is a complex thing, and sometimes requires a people to invent a narrative of the past that legitimises the nation, a process that can be called nationalism.[4] This viewpoint is developed with numerous examples by Benedict Anderson in *Imagined Communities, Reflections on the Origins and Spread of Nationalism*. Anderson has argued, persuasively and influentially, that nations emerged only after three beliefs were weakened:

- that elite languages (like Latin) offered unique access to truth about existence;
- that society was naturally organised around leaders who ruled through divine dispensation;
- and that the origins of the world and of humankind were essentially identical.

It was economic change, scientific discoveries and a revolution in communications that broke down the old beliefs and allowed enterprises to emerge.[5]

Some have associated nationhood or nationalism with a common language, but experience says that is neither necessary nor sufficient. Germans had a common language across many states for a long time before unification in 1871 (actually occurring progressively over the 19th century) but could not begin to achieve the advantage they eventually did because they did not feel and act like a nation. The Swiss have had a strong national identity (and achieved considerable competitive advantage in the post-World War II period) despite speaking four languages in a very small country, albeit one where German is the language of more than 60 per cent of the population.

Even the United States, having common language and central government, struggled with competing views of its nationhood up to the wrenching effects of the Civil War in the 1860s and for a very long time in the aftermath of that war, much of which was about the role of individual states versus that of the 'United States'. In 1860 the US, with all its natural resources, had a lower GDP per capita than the UK, but by 1913 these positions were reversed. It was necessary to throw off the divisions between slave states and free states if a single national identity was to be achieved. Arguably it still took both world wars to bring the country together as a nation and lay the foundation for the competitive advantage the US achieved in the 20th century, though many of the fundamentals for this success were already in place. By contrast, Britain at the beginning of the Industrial Revolution felt very much like a nation, in terms of how people identified, where their loyalty lay, and the direction of travel on social issues (though admittedly it was a long road to be travelled). In other ways, of course, Britain did not fit a traditional nationalism, because the English, Scots, Welsh and Irish saw themselves as distinct nations. Germany and Japan also developed greatly enhanced feelings of nationhood in the last few decades of the 19th century.

In Chapter 5 the role of paper as a key material for information dispersal is discussed. Benedict Anderson gives many examples of how the newspaper, which he refers to as 'print-capitalism', plays a central role in the development of nationalism. Indeed, in countries like Switzerland, where the everyday German spoken is several dialects and many citizens were not particularly good at understanding High German, the rise of nationalism was delayed by the absence of a newspaper everyone could understand. This will be discussed further in Chapter 5.

## FUNDAMENTAL CHARACTERISTICS OF A COMPETITIVE NATION

What are the 'fundamentals' about a society and how it is governed that enable competitive advantage through enterprise formation and growth? It is best to preface what follows by saying that no nation is perfect, far from it, and probably none ever will be in terms of the qualities being discussed. But this is about competition, not perfection. What a nation needs to be is better than those with whom it is competing across a range of civil society measures: governance; the rights of individuals to profit from their creativity and determination; and incentives for workers to become more productive.

David Landes[6] took a view on what was the ideal case for a society that will grow, develop and thrive. Such a society, said Landes, would be one that:

- *Knew how to operate, manage, and build the instruments of production and to create, adapt, and master new techniques on the technological frontier.*
- *Was able to impart this knowledge and know-how to the young, whether by formal education or apprenticeship training.*

- *Chose people for jobs by competence and relative merit; promoted and demoted on the basis of performance.*
- *Afforded opportunity to individual or collective enterprise; encouraged initiative, competition, and emulation.*
- *Allowed people to enjoy and employ the fruits of their labour and enterprise.*

In Chapters 3 to 8 there will be numerous examples of these societal attributes enabling success.

There are, Landes said, corollaries that are implied, for example, gender equality, lack of discrimination based on irrelevant criteria such as race, religion, etc., and a strong preference for science and rationality over magic and superstition. Societies have certainly succeeded without being anywhere close to perfect on these corollaries (especially the first two), but in most cases they were better than those with whom they were competing and directionally moving towards these principles rather than away from them.

If these are the larger goals of a society, they imply a set of political and social institutions that will work towards achieving these goals, for example (again quoting Landes):

1. *Secure rights of private property, the better to encourage saving and investment.*
2. *Secure rights of personal liberty — secure them against both the abuses of tyranny and private disorder (crime and corruption).*
3. *Enforce rights of contract, explicit and implicit.*
4. *Provide stable government, not necessarily democratic, but itself governed by publicly known rules (a government of laws rather than men). If democratic, that is based on periodic elections, the majority wins but does not violate the rights of the losers; while the losers accept their loss and look forward to another turn at the polls.*
5. *Provide responsive government, one that will hear complaint and make redress.*
6. *Provide honest government, such that economic actors are not moved to seek advantage and privilege inside or outside the marketplace. In economic jargon, there should be no rents to favour and position.*
7. *Provide moderate, efficient, ungreedy government. The effect should be to hold taxes down, reduce government's claim on the social surplus, and avoid privilege.*[7]

Collectively, the first five points, the corollaries, and these seven points can be referred to as the 'Landes principles'. Looking back to Chapter 1, these principles can be seen as an early articulation of certain key aspects of sustainable development.

An additional perspective on the characteristics necessary for national or regional competitive advantage comes from Niall Ferguson,[8] and is worth enumerating because it both reinforces and adds to the points already made. In considering why the US and some European countries achieved competitive advantage, he enumerates six factors:

1. *Competition – within countries, be they monarchies or republics, there were multiple corporate enterprises that competed with one another.*
2. *Science – the major scientific breakthroughs from the 17th century onwards in mathematics, physics, chemistry and biology occurring in these countries.*
3. *Rule of Law – and representative government, particularly as it emerged in the English-speaking world, including protection of property rights.*
4. *Modern Medicine – controlling the spread of tropical diseases; one can sometimes forget that such diseases were prevalent even in the early United States.*
5. *Vibrant Consumer Society – so that the Industrial Revolution took place where there was both a supply of productivity-enhancing technologies and a demand for more, better, and cheaper goods.*
6. *Work Ethic – such that structured work patterns were rewarded and led to higher savings, sustained capital accumulation by families, and feedback into the consumer society, including housing.*

Overlaying these factors on the systems diagrams of Chapter 1 is illuminating – where there is a strong scientific base, the processes for conversion of raw materials into products emerge. Where there is a vibrant consumer society, the increases in demand occur that drive down costs, as well as open up markets for new products using the same materials, leading to strong positive feedback. The Landes principle of providing 'opportunity to individual or collective enterprise; encouraging initiative' encapsulates the crucial role of entrepreneurship in an economically successful nation, while the strong work ethic allows for increases in productivity, again with positive feedback.

## CENTRAL ROLE OF THE EMERGENT MIDDLE CLASS

*The natural effort of every individual to better his own condition, when suffered to exert itself with freedom and security, is so powerful a principle, that it is alone, and without any assistance, not only capable of carrying on the society to wealth and prosperity, but of surmounting a hundred impertinent obstructions with which the folly of human law too often encumbers its operations... In Great Britain industry is perfectly secure; and though it is far from being perfectly free, it is as free or freer than in any other part of Europe.*

Adam Smith, *Wealth of Nations*[9]

It is difficult in practice for a nation that still has massive class distinctions, unresponsive government and a system that is not meritocratic to be explicit about its long-term goals. This was Britain in the 18th century. But viewed through a long enough historical lens one can see an inexorable direction of travel, while other nations with which it was competing did not show the same trend. One of the most important ways this direction manifests itself is in the growth of a middle class. There are many competing definitions of middle class; for the purposes of this discussion it is sufficient to say that once a household has sufficient income beyond their basic needs of food, clothing and shelter to spend on other things – that is 'disposable income' – they have entered the middle class.

Using this operational definition, it is easy to see the importance of the middle class to domestic markets for manufactured materials, which helps drive industries producing novel goods forward to scale, even before they access export markets for their products. Clothing is a good example, discussed in more detail in Chapter 4. The poor will spend as little as possible on clothing, prioritising food and shelter. The upper classes want clothing, often custom-made, that is distinctive to their standing in society. But for a middle-class person, good quality, standardised (but not uniform, hence stylish) clothing becomes important. A civil servant or an office worker needs to dress appropriately to their position at work. A middle-class child, teenager or university student (because access to tertiary education is associated with the middle class) must have suitable clothing for both school and leisure. The increasing scale of this market led to standardised sizes, mass manufacture and reduced costs following the dynamics shown in the systems diagrams introduced in Chapter 1. The Industrial Revolution provided, in waves, similar changes for energy, iron and power. As such, it completely transformed society in Britain and yielded competitive advantage that endured for centuries.

The middle class also places demands on society that reinforce the Landes principles and corollaries outlined above. They have the where-withal to finance services such as public education, transport and health care, and are willing to pay taxes to get these services. At the same time, they hold government accountable for the quality of the provision, as well as the integrity of the processes that were used to spend the money they provided. Because they engage in buying and selling houses, banking, investing and other commercial transactions, they insist on the enforcement of contracts, on a fair system of regulation and on the rule of law impartially enforced. A growing middle class thus feeds forward into creation of new industries, and feeds back into a stronger commitment to the principles that led to its growth. All of which begs the question of how to increase the middle class.

## RESOURCES CREATED BY THE COMPETITIVE NATION

The answer to that question requires a step back to view the economy not just from the Landes principles for a society, but from the point of view of resources, broadly defined. Economists considering competitive-ness in world trade have typically spoken of these in terms of 'factors'. Traditionally, factors were grouped as land, labour and capital, perhaps more suited to an agrarian than an industrial economy. In a modern classic book revisiting this approach, Michael Porter[10] grouped them as:

- *Human resources: the quantity, skills, and cost of personnel (including management), considering standard working hours and work ethic...*
- *Physical resources: the abundance, quality, accessibility, and cost of the nation's land, water, mineral or timber deposits, hydroelectric power sources, fishing grounds, and other physical traits. Climatic conditions can be viewed as part of a nation's physical resources... The time zone of a nation relative*

*to other nations may also be significant in a world of instantaneous global communications (e.g. London between the US and Japan).*

- *Knowledge resources: the nation's stock of scientific, technical and market knowledge bearing on goods and services. Knowledge resources reside in universities, government research institutes, private research facilities, … business and scientific literature, … and other sources.*

- *Capital resources: the amount and cost of capital available to finance industry (debt, equity, venture capital, etc.).*

- *Infrastructure: the type, quality, and user cost of infrastructure available that affects competition, including the transportation system, the communications system … payments or funds transfer, health care … also includes housing stock and cultural institutions.*

These may be referred to as Porter's factors. To these should be added the preservation of resources through the principles of sustainable development, for example by living within environmental limits as discussed in Chapter 1, for the very reason that sustainable competitive advantage requires a country to be superior to its competitors in this aspect as well as the others.

This categorisation of factors is strikingly different from the traditional emphasis on physical or natural resources and builds on assertions made in Chapter 1. While a nation may be better or less well endowed with the physical resources of minerals, arable land, timber, etc., this is, in general, not the most significant determinant of competitiveness. Examples will come up repeatedly in this book of nations being very successful in industries where the key physical resource was completely lacking. The British dominated the cotton textile industry without having the climatic conditions to grow any cotton at all. The US, Britain and France dominated the tyre industry without growing any rubber. Porter discusses the success of Switzerland in a number of industries, showing how the factors other than physical resources are critical. By contrast the southern US, despite having ideal growing conditions for cotton, did not achieve competitive advantage in cotton textiles, lacking many of the other Landes attributes.[11] Still, the key physical resources are important, and countries that have them have often been exploited as a result.

The difference between the physical resources and the other factors is that each of the others is something that a country creates, nurtures and protects through the societal attributes and governance structures rather than being a natural endowment, though of course factors such as arable land must also be cared for in a sustainable way. These created factors link back very directly to Landes' and Ferguson's lists, for example affording opportunity to individual or collective enterprise; encouraging initiative, competition and emulation; allowing people to enjoy and employ the fruits of their labour and enterprise. In practically every materials story described in subsequent chapters, a strong system of protection of intellectual property plays a role in the realisation of this ability for people to enjoy the fruits of their enterprise. The factor of scientific resources (and the engineering to realise these in practice at world scale) is explicit in Ferguson's list and is enabled by almost all of Landes' five societal characteristics. Likewise, Landes' warning against discrimination based on race or religion led to nations with strong competitive positions (especially Germany) losing this advantage by expelling scientists on precisely that basis, while those that were welcoming to such refugees benefited. Before one even gets to having a science and engineering resource, there are the fundamentals of education. While literacy as a materials story is discussed explicitly in Chapter 5, it is implicit in every case of competitive advantage in this book.

## THE CHANGED NATURE OF WORK – PRODUCTIVITY

What happened to work when it left the home or farm and moved to the factory so that manufactured materials could be produced at scale? Of course, many horrible and exploitative things happened. The word disruptive is often used these days to describe the effects of a new technology on work, but few changes could have been more disruptive than this one. There were terrible abuses and bad working conditions, certainly not universally but widespread. People crowded into cities from villages and the countryside, and the infrastructure was not there to support the rapid increase in population.

Despite all those problems, there was a structural change in the nature of work. The working day, and week, became defined as having a beginning and an end, which it never had before for most of the population. Wages were paid at a fixed time. In some cases, employers took on responsibility for health care and other benefits, sometimes including housing, for their workers. Eventually, there was provision for retirement, a new concept, although life expectancy meant this was, on average, a relatively short period.

There is a consequence to this structure of work: it is inherently meritocratic if a company is to be successful. One employee in a team of five to ten will stand out and become a team leader. He or she will then become part of the hierarchy in the factory, associating with others in a similar role. Construction work will likewise be structured into teams with supervisors or foremen. These team leaders and supervisors need to be chosen based on merit rather than family connections or length of service. The factory will require a number of services, such as accounting and associated bookkeeping. From among the most literate and hopefully the more capable, young people will be trained to take on these responsibilities. Contracts need to be drawn, patents written, legal work of all sorts undertaken. The people to do all of this did not come from the upper classes moving down but emerged from the lower classes moving up. In this way it was inevitable that a middle class was created and grew.

The changed nature of work also meant that for many workers they only produced part of a product rather than the whole thing. This was true even in the 1800s, and became even more so with the development of the high-speed assembly lines of 20th-century industry. This was one of the first aspects that Karl Marx found so detrimental to the life of the workers – that they could not see a product from beginning to end in their daily efforts.

The final critical point about this organisation of work is that it must have a bias towards increased productivity. Productivity is most simply defined as output per worker, though the literature has many more refined definitions and measures. Productivity must be achieved by the manufacturing process being done more efficiently and effectively. This involves both the workers and the machinery they use. One logical path says that workers should be biased against increased productivity, because it would mean that the same output can be delivered with fewer workers. Increase everyone's productivity by 10 per cent and one out of every 10 workers is not needed. What works against this destructive train of thought is a recognition by the workers that if they are more productive, the company will be able to grow faster than competitors, be more successful, and more workers will be needed. Moreover, the company will be able to increase the pay of the workers because it is more profitable. It is one of the core tasks of the leadership of any business to demonstrate through their actions that it is in the interest of the workers to be more productive.

Productivity gains need to be an example of the reinforcing mechanism in the systems analysis in Chapter 1. If this is done properly, more workers have more income to spend beyond their basic needs. It took several generations of mechanisation and industrial revolution for this to happen. In some cases, perhaps in many cases, it required unionisation of the workforce and strike action for workers to get their share in the fruits of increased productivity. Such struggles continue through to today, and only in hindsight is the direction of travel clear. By the middle of the 20th century auto workers were middle class. So were the workers at the successful companies in the vast supply chain that underpinned the industry.[12]

As mentioned earlier, the middle class demands and is willing to pay, through taxes, for better public education. They do this because of the aspirations they have for their children to be fit for positions higher in the middle class, that is, they are willing to sacrifice some of their own disposable income to invest in the next generation.

Is productivity growth an unalloyed good? Certainly not. To the extent that, as described, it is productivity growth that leads to a bigger economy producing and consuming more goods, hence producing more waste, driving environmental destruction – including as manifested in climate change – productivity is not something that we must use as a measure of societal progress. (Systems analysis shows that this can correct itself to some extent, in that climate change can prove to be a negative driver on productivity.[13] This has been taken up by Aghion et al[16] and forms the core of their argument for a carbon tax.) The negative view of productivity and growth goes back more than 50 years to the publication of *Limits to Growth*,[14] which asserted that economic and population growth cannot continue along the exponential curves they had been on because of finite resources, including environmental capacity. While heavily criticised at the time of its publication, many of the predictions it made have been validated. However, much has also changed, with slower population growth correlated with increased educational levels, particularly for women. Technological developments have also opened the possibility of abating the impacts of growth, if societies choose to do so. Some of these possibilities, for example recycling of paper and aluminium, have already had major impact on resource demands. As described in the systems diagrams of Chapter 1, such actions are themselves included under the rubric of productivity growth. Nonetheless, a modern competitive society must view productivity in the light of the impacts of growth.

All this has been about the role of materials-based enterprises being founded, nurtured and growing to such a scale that they achieve significance well beyond the nation's borders. They change society and provide wealth that ultimately allows that nation to reinforce its advantage over others. To achieve this many of the Landes principles must be in place. Countries that ignored or, worse, consciously violated these principles, have consistently lost competitive position over decades and sometimes over centuries.

## SCIENCE, ENGINEERING AND INVENTION

This book is about advantage related to materials, and as described right at the start of Chapter 1 the first key step is the scientific research and the inventions to take a raw material, such as an ore, a tree, a rubber plant or a sheep, and turn it into aluminium cable, paper, a tyre or a woollen jumper. After many centuries of technological development, there is some understanding of the routes by which this happens. Sometimes it is systematic, studying basic processes, developing an understanding, and from this finding the route to the desired product. Wallace Carothers and his colleagues at DuPont were following such an approach in polymer science to create synthetic fibres in the 1930s. Likewise, although they did not understand the basic science of genetics, sheep breeders did work very systematically over several centuries to increase wool production from their herds. The Wright Brothers studied everything that was known about aerodynamics and advanced the science of it themselves in their pursuit of powered flight. Charles Hall used the accumulated science of electrochemistry in the late 1800s to produce his process for making aluminium.

Sometimes developments are the result of accidents. Charles Goodyear discovered the process that became known as the vulcanisation of rubber as the result of a spill of sulfur onto a heated pool of latex. Teflon was accidentally discovered by Roy Plunkett at DuPont. Plunkett's first assignment as a junior scientist at DuPont was researching new chlorofluorocarbon refrigerants. Plunkett had produced a hundred pounds of tetrafluoroethylene gas (TFE) and stored it in small cylinders at dry-ice temperatures before doing the next chemical step of adding chlorine. When he and his helper retrieved a cylinder for use, none of the gas came out – yet the cylinder weighed the same as before. They opened it and found a white powder, which Plunkett had the presence of mind to characterise for properties other than refrigeration potential. He found the substance to be heat resistant and chemically inert and to have very low surface friction so that most other substances would not adhere to it. Plunkett realised that against the predictions of polymer science of the day, TFE had polymerised to produce this substance – later named Teflon – with these potentially useful characteristics.[15] While the discovery was accidental, Plunkett always maintained that it was his strong scientific training as a PhD chemist that led him to realise its significance. All this scientific expertise is there because the nation has created the knowledge resource as shown in the Porter factor list.

Invention is different from science.[16] In science one does experiments that build on everything that has gone before, and in most cases the result is an incremental advance in knowledge. It is only when an experiment yields something unexpected, and this is rare, that the science takes a different turn. Scientific research is disclosed in the literature of that discipline. In general, basic science is not trying to produce something useful, although scientific research in industrial laboratories is often in support of a commercial goal. By contrast, for an invention to be patentable it must be novel, useful, non-obvious and not previously disclosed.

Inventors thus look at the world in a completely different way from scientists, indeed from almost everyone, even though scientists such as those at DuPont and IBM did receive many patents for the useful outputs of their research. After something is invented and brought to commercial use there is often a lot of science put in behind it to understand why it works, and that understanding underpins the multiple stages of evolution of the original invention.

Materials discovered in laboratories (in the most general sense: for example, in this context a sheep farm can be a laboratory) become important sources of competitive advantage when they are produced at large scale, and this requires both quantity and quality of engineering. To take a novel polymer such as Teflon from a few grams of white powder inside a cylinder to a product selling 250,000 tonnes per year requires chemical, mechanical, electrical and controls engineers. This is why the human resources Porter factor is so crucial to national competitive advantage. It cannot be created instantly to serve a particular need but must be built up through universities and other institutions over decades. Germany moved ahead of the United Kingdom competitively because it realised that systematic education of a respected corps of engineers would be important to industrial success.

The Morrill Acts in the US in 1862 and 1890, giving land to the states to set up colleges for teaching and research in 'agriculture and the mechanical arts', were crucial to the education of the large quantity of engineers that created the 20th-century competitive advantage of the country. The establishment of the National Academy of Sciences by the US Congress in 1863 to be an organisation of the nation's most distinguished scientists created a body that the government could turn to for impartial advice. The charter of the National Academy provided that it would respond 'whenever called upon by any department or agency of the government, to investigate, examine, experiment, and report upon any subject of science or art.' In the UK the Royal Society has fulfilled a similar role since about 1800. It is interesting that engineering excellence and the need for an advisory role from the most outstanding members of the engineering profession was only recognised by establishment of the US National Academy of

Engineering in 1964 and in the UK the Royal Academy of Engineering in 1976.

## THE ROLE OF GOVERNMENT

What is the role of national government in all of this? Can government decide that the country will achieve excellence in one or more industrial sectors and make it happen? Can government protect industries from external competition, so enabling them to become more competitive themselves? Or are there only more generic things that government can put in place to foster competitive advantage? The broad spectrum of government interventions to help industry achieve competitive advantage is known as industrial policy. The idea goes back many centuries, and examples will be seen in this book in areas such as clothing, where some governments tried to criminalise individuals with expertise sharing it beyond their borders. Most of these policies/rules ended in failure, though industrial policy itself is not inherently a failed idea, indeed it has had dramatic successes right up to the present day.

One way of looking at the role of government is to put in place those of the Porter factors that are within its capability to deliver, and not easily established by individual companies or industries. Examples (some of these may be at local or regional level rather than national) include infrastructure such as transportation, the regulations in support of an efficient and trustworthy banking system, a high-quality educational system at all levels, from primary through university, support that ensures a more than adequate number of scientists and engineers, support for scientific research of the highest quality (including self- or external assessment of the quality of that research in the global context), and a robust system for protection of intellectual property. In every one of the chapters that follow there are examples of factor creation of this sort being crucial to competitive advantage.

Government also plays a role as a regulator, and this can also be at national or more local levels. Can government regulation ever enhance national competitive advantage? Yes, and in many different ways. Michael Porter and Claas van der Linde[17] discussed one example of this in considering environmental regulation. This is a particularly good case to consider because it is often asserted that if one country places such regulations on its companies they are competitively disadvantaged. This complaint was a persistent theme articulated by the Trump presidency in the US. Done properly, environmental regulation can make the industry of a nation more competitive globally. Regulators must set the bar high but give companies time to comply. This accomplishes two things: incremental change (tinkering) will not do the job, and a sufficiently long interval encourages innovation. Government regulation must also focus on desired outcomes, not on the processes to achieve those outcomes. In so doing, government encourages innovative approaches and rewards invention. Government is competent to determine what the important outcomes are, while industry is most competent to figure out how to achieve them. But the most crucial point of this is a reiteration of the importance of living according to the principles of sustainable development discussed in Chapter 1 – living within environmental limits. Ultimately, when a nation has high standards of environmental protection it is more prosperous internally (for example, it has lower health care bills) and its companies are more attractive to customers internationally.

A seemingly fairly benign form of government intervention and regulation is in setting standards for all sorts of manufactured materials. This starts with weights and measures, which have become global but at one time were very much a national preserve. In the US, by 1850 the Coast Survey Office would issue to the capital of every state, in addition to a set of standard weights from 50 lbs down to 1/10,000 of an ounce, a measure of a yard, liquid measures from a gallon down to a half pint, and a half bushel, along with a set of three balances.[18] This is fundamental for business, but it is only the beginning. There must be standards for nuts and bolts, hose fittings, brick sizes, and on and on. This is 'seemingly fairly benign' because while all manufacturers can agree the desirability of having such standards, some will benefit by their particular nuts and bolts being chosen as the standard, while others will have to retool.[19] Standards of these sorts, across the full range of manufactured materials, are characteristic of countries that lead in manufacturing, although many of them only appeared in the early 20th century. Even such a mundane thing as standards can have positive feedback, because the leading country that establishes these standards imposes them on others, forcing them to incur the costs required to comply.

Creating, managing and enforcing the protections offered by patents, copyrights and trademarks is also a role for government. A robust system of intellectual property protection gives inventors, entrepreneurs and investors confidence. For materials, since the early 1800s, this has been crucial, and many examples of key patents are given in Chapters 3 to 8. No patent system is perfect. The examiners have a difficult job in trying to understand submissions, prior art, lack of obviousness and other factors. Some patents will be granted that should not have been approved, but where the system works, challenges to these are heard in a legal system that has a high level of competence and operates impartially. By contrast, where a government does not uphold intellectual property rights, or in some cases encourages infringement to obtain competitive advantage for its own companies, in the long term there will be a lack of investment, particularly in manufacturing.

Government's regulatory role extends to discouraging a variety of bad behaviours. Among these are legislation and its enforcement to stop all forms of anti-competitive practices. Mergers to eliminate competition and create dominant market position, cartels to fix pricing, collusion between companies against the interests of customers and suppliers – all of these are eventually detrimental to national competitiveness.[20]

Not all regulation is at a national level. Cities and city regions usually have control over many aspects of planning, and successful cities use this to create advantage. This may be in zoning decisions for residential, industrial, commercial, retail, etc.; it can be in encouragement or discouragement of higher density through commercial skyscrapers, tall residential buildings, mixed-use buildings; setting standards for energy efficiency for new buildings; in many cases use of tax policy to attract business or support particular development – all this and more is in the power of cities that want to gain competitive advantage.

Where governments fail is when they do things that they are not competent to do, certainly where they are far less competent than private companies. Examples include protecting companies from domestic competition, putting up barriers to trade (even when companies urge them to do these things), deciding that most/all research must be applied rather than basic, and then selecting those areas of application that it is sure will be the winning ones. Perhaps the biggest failure in terms of building highly competitive industries or sectors is nationalisation. Once an industry is owned by the government there is no incentive, not even a mechanism, to increase productivity, to control costs, modernise or to innovate. Certainly, there are a few exceptions, but in almost every case these are highly profitable businesses that government allows to run at arm's length so that they provide cash for other governmental uses, for example Saudi Aramco or the Panama Canal. All this not to imply that the market does things perfectly – far from it. Indeed, businesses often become successful because they recognise market imperfections that their competitors do not appreciate. Rather it is to say that government has vital roles to play, that in successful nations governments become exceptionally skilled at carrying out those functions, or at least more skilful than other nations, and they do not overstep into areas for which they are not well equipped.

## THE CORPORATION – LIMITING LIABILITY

The corporation as an entity, with executives and a board of directors, is so commonplace today throughout the world that it is worth remembering that key aspects of this legal construct are only about 150 years old. Corporations certainly existed long before that, for example the East India Company and the Dutch East India Company, but they could only be established by acts of government.

Aside from the few corporations that were established by governments, business was historically conducted as sole proprietorship, partnerships, or some form of trust. For all of these there was a crucial issue: when things go wrong, whose assets are at risk? For capital markets, as described in the previous section, to operate in support of growing a new industry, the investors must know that, assuming laws have been followed, their personal assets are not being pledged to pay the debts of the company.

The key features of a corporation are these:

- limitation of liability, so the most that an investor (shareholder) can lose is the amount that they invested;
- transferability of shares, that is, one shareholder can sell their shares to another party, either in a private transaction or through public markets (stock exchanges) without the entire business arrangement having to be reconstituted, as would be the case with a partnership;
- for legal purposes, the corporation is treated as a 'person', that is, it can sue and be sued, it can enter into contracts, and own property in its own name;
- exists for an indefinite period of time.

The legislation that allowed this to happen was put in place in the 19th century in Britain, the US, France and Germany, with its scope and limitations being clarified by a series of legal decisions. The US was, and remains, a special case because the laws governing corporations were deemed to be a power of the individual states, while the constitution proscribes restrictions on interstate commerce. Over time, the small state of Delaware, which was one of the first to put such laws on its books, as well as establish courts and other mechanisms to regulate them, has become the single most popular place for establishing a US corporation.

How such corporations present their accounts has also become a subject of standardisation, starting with double-entry bookkeeping[21] and progressing to modern standards like GAAP (Generally Accepted Accounting Principles). This means that an investor can compare the performance of companies against a common standard of reporting, auditors can check compliance with the principles, and regulators can assess whether the standard has been met.

The emergence of legislation establishing corporations, and the court decisions that supported the idea, gave the four pioneering countries substantial competitive advantage in mobilising the capital needed for vast industrial growth that occurred in the late 19th and early 20th century, as will be seen in subsequent discussion of buildings and infrastructure, transport and electrification.

## THE ROLE OF CAPITAL MARKETS

At every stage of the transformation of a raw material into a range of manufactured products available at low cost and large scale there is a requirement for money to be injected. The money for research will usually come from both government and corporations. There is no single preferred model for this – probably hybrid models where there are multiple sources of research funding are the most robust to sustaining innovation through economic cycles.

A research achievement that emerges from the laboratory as a potential commercial enterprise needs patient capital, because where manufacturing is involved, it takes time to convert a laboratory result into an industrial process, even at small scale. Moreover, there will be failures – lots of them. Sometimes these are for technical reasons – the laboratory result did not show up problems that occur when materials are made in quantity; some- times these are due to not having the competencies in the people selected to take the project to the next stage; sometimes investors decide that they have reached their limit and no new investors can be found, even for a good idea that will ultimately succeed. In this book are the stories of successful entrepreneurs and the capital that backed them. But there were hundreds of failures in making automobiles, canned food, machines to automate textile production, aluminium purification and of course powered flight. The failures are usually lost to history. Governments can play a role in providing incentives (usually in the form of tax relief) for those who risk investing in early-stage ventures, and sophisticated capital markets have learned to be tolerant of failure, indeed, to expect it.

The amount of money required to take an idea and convert it to a research achievement is on the order of $1 million, sometimes less. To build a first plant that makes this into a manufacturing process producing a product requires 10 to 50 times as much capital from venture inves- tors. To then take this into a large-scale industrial process may require a further 10- to 100-fold increase in the amount of capital required. This is where the role of public and private capital markets come in. The risk of failure decreases, so the investment is suitable for those with a lower ability to tolerate risk of failure, such as a pension fund. At the same time, the expected returns are lower, though they may be substantial over time.

The presence and efficient operation of private and public capital markets are an essential component of a nation that is going to gain competitive advantage and sustain it over time. Rewarding success, tolerating failure, funding industrialisation to scales of global significance – all are characteristics the markets must embody. The amounts of capital vary widely, and they have certainly changed over time, but the existence of these markets and their essential features endure.

## ADVANTAGED CITIES AND CITY REGIONS

Thus far all this discussion has been about nations achieving competitive advantage over other nations. Especially in large countries, sometimes even in much smaller ones, certain cities or city regions excel compared to others in the same country. Porter[4] discusses the idea of clusters, sometimes referring to all the related parts of an industry being clustered in a particular country (for example, German printing machinery, papermaking and inks) as well as how certain industries grow up around a city or city region, such as the US automotive industry in Detroit, tyres in Akron, ceramic tiles in the Emilia-Romagna region of Italy, textiles in Manchester, financial centres such as London and New York, oil refining in Houston, Rotterdam and Singapore, and in the 21st century digital technologies in Silicon Valley and Seattle.

The roles of government as regulator at city level have already been mentioned. Additionally, some governments at the city level saw the importance of infrastructure such as public transport, bridges and tunnels and pursued their construction earlier and more vigorously than others, often securing national government support to underwrite this development. In this way taxpayers from across the nation subsidised selected cities to become advantaged over others.

There is a line of thinking that says that the competitive advantage of cities secured in this way – whether by infrastructure, tax benefits or planning policies – is not sustainable, because it has no inherent intellectual property; any other city can copy it. This has been proven wrong many times. It is of course wrong for clusters, because there are advantages to being part of one in a particular industry or in a sector such as technology venture capital – labour, exchange of ideas, supplier presence – that cannot be easily duplicated. But it is also true for financial centres such as New York and London, because there is a less well-defined feeling of 'needing to be a part of it' that comes with such agglomerations. How technology facilitation for working from home (in part virally imposed) affects this in the 21st century remains to be seen, but it is too early to write off the attraction that most of the workforce sees in the hospitality and cultural amenities that are part of big cities. The vulnerability of clusters is not in their being copied but in their industries declining or being supplanted by newer technologies or more productive manufacturing processes with which they have not kept pace. Cities like Detroit, Cleveland and Akron in the US went from being centres of great affluence and influence to 'the rust belt' as competitors took their previous industrial leadership away.

Likewise, infrastructure of cities and city regions must be kept up to date. London and New York had world-leading public transport for many decades that enabled their growth and prosperity. Lack of ongoing investment turned this from a resource advantage to a disadvantage when they failed to maintain and upgrade the systems through decades of underinvestment.

## ENTREPRENEURIAL ACTIVITY

*Entrepreneur Noun [Fr] a person who sets up a business or businesses, taking on financial risks in the hope of profit; one who undertakes an enterprise, especially a commercial one, often at personal financial risk.*

An outcome of a vibrant culture of science/engineering/invention, a progressive government policy, active capital markets and advantaged clusters of activity is entrepreneurship. This is when a person or persons with an idea create a firm to realise that idea, aggregating labour and capital to produce goods or provide services.

There is a huge variation in entrepreneurial activity between countries today, and this has been true over the many centuries of materials development that will be discussed in Chapters 3 to 8. An attempt to measure this is the Global Entrepreneurship and Development Index (GEDI)[22], which shows the top ten countries as

| 1 | United States | 83.6 |
|---|---|---|
| 2 | Switzerland | 80.4 |
| 3 | Canada | 79.2 |
| 4 | United Kingdom | 77.8 |
| 5 | Australia | 75.5 |
| 6 | Denmark | 74.3 |
| 7 | Iceland | 74.2 |
| 8 | Ireland | 73.7 |
| 9 | Sweden | 73.1 |
| 10 | France | 68.5 |

Of the bottom ten countries (of 137 ranked), nine are in Africa, with ratings on the same scale of 9 to 13.

GEDI also recognises that within a country there are differences in entrepreneurial activity and publishes regional indices as well.

One way of characterising the aspects of this activity is:[23]

| Attitudes | > | Opportunity perception<br>Start-up skills<br>Non-fear of failure<br>Networking<br>Cultural support | > | Key focus for<br>factor-driven<br>economies |
|---|---|---|---|---|
| Abilities | > | Competition<br>Quality of human resources<br>Tech sector opportunity start-up | > | Key focus for<br>efficiency-driven<br>economies |
| Aspirations | > | Risk capital<br>Internationalisation<br>High growth process innovation<br>Product innovation | > | Key focus for<br>innovation-driven<br>economies |

These attitudes, abilities and aspirations are shaped, and differentiated, by the educational and cultural milieu in which people grow up. How are they educated, what values do their parents communicate to them, especially in terms of the balance between risk and job security, how do they see their ability to succeed in society? Entrepreneurship happens in the systemic environment of resources, infrastructure, government regulation and incentives, sophistication and risk appetite of capital markets, and awareness of opportunities. Without it there would be few new jobs and little opportunity for a society to achieve competitive advantage.

The list of the top ten countries in the GEDI is instructive. It includes large countries and small. Their governmental and social policies span a range from low to high levels of intervention and social support, lower and higher tax regimes, labour policies that give employers great freedom to reduce workforce size to those that restrict this severely. Some of these countries allow great wealth opportunities to successful entrepreneurs, particularly through tax incentives on capital gains. They represent a range of inequality as measured by the GINI coefficient, from a low of 26.1 per cent for Iceland to a high of 41.4 per cent for the US, although they cluster around 30–35 per cent.[24] Some, such as Sweden, frown on share option schemes and make them difficult to implement, while in the US and UK such schemes are nearly universal in technology companies. But all of them have strong educational systems, a commitment to science and engineering, and legal systems that protect intellectual property. So there are commonalities, to be sure, but it is important to recognise there is not a single model that is required for a vibrant entrepreneurial culture.

## MERITOCRACY

*Once upon a time, the distribution of power and privilege was determined by birth. Now it is determined by merit. And that, in a nutshell, is the history of the long 20th century.*
Stefan Collini in *London Review of Books*[25]

Slavery, employment and advancement on anything other than merit, discrimination based on race, gender or religion: these are corrosive practices. Everyone in a society should feel that they or their children have the possibility to improve their position during their lifetime based on their own work and achievements.

This means there must be access to education of quality available to all. Education is the start. In our times South Korea has shown this more clearly than others, starting in 1959 with a commitment to free, high-quality primary education and continuously improving the educational system through the grades and decades. Today South Korea's educational system and achievements of its students rank among the top five in the world across a range of subjects. And there has been, in parallel, dramatic and continuous growth in GDP per capita.

The name given to the principle that advancement in society is based on one's own achievements rather than gender, class, parental wealth, or any other criterion, is meritocracy. Earlier in this chapter, discussing the changed nature of work, a basic form of meritocracy in the workplace was described. But the bigger dogma, of which millions across societies as diverse as the US, Western Europe, China and Japan were convinced during the 20th century, is that this would be true across society in general. Superficially simple, meritocracy actually is a very complex and troubled subject.[26] Do the best jobs in society go to the most intelligent or to the hardest working, perhaps the most intelligent of those who work the hardest? In any case, what are the best jobs, are they the same as the ones that pay the most? Does everyone have access to the same quality of education, not just at university level but as early as pre-school?

Each of the nations considered in this book are different and were in very different positions with regard to meritocratic opportunity at the time when materials became sources of competitive advantage. Certainly in Western Europe and the United States there was a great expansion of economic mobility based on merit during the 20th century, at least until the late 1970s. Evidence of the last 50 years indicates that this is less the case now than it used to be. Why might that be? While the wealthiest 0.1 per cent of society have always had access to special educational and other resources for their children, and the least wealthy, say the lowest 50–90 per cent in terms of income, have been disadvantaged in this regard in many but not all societies, there is a view that persistent inequality in many countries is a result of the wealth and behaviour of the 10–20 per cent most affluent.[27] Their children often attend very selective pre-schools, get special tutoring, have a diverse set of travel experiences, get summer jobs and internships based on familial connections. As a result, the link between family income and intergenerational mobility (the opportunity of children to advance) has become stronger for this group. Put another way, how strong is the correlation of children's income with that of their parents? A measure used by economists, Intergenerational Earnings Elasticity (IGE), would be zero if there is no correlation (perfectly meritocratic society) and one if there is perfect correlation. In the US the IGE has risen from about 0.3 50 years ago to nearly 0.5 today. The UK, Italy and France are similar to the US, while Finland, Norway and Denmark are about 0.2. This is a significant difference in equality of opportunity.

Overall, the world is becoming a more equitable place. The distribution of per capita income has become both broader and peaks at a higher level.[28] Nonetheless, the drift away from a meritocracy evidenced by the wealth accumulation and behaviours of the 'upper middle class' in certain countries could be a negative factor in their future competitive position.

## FINAL WORDS – FOR NOW

The characteristics that a nation must have if it is to develop competitive advantage through materials – indeed through any industrial process – are clear. There is an ideal, and no nation has ever achieved this ideal, but advantage requires a nation to be better than those with whom it is competing, and the direction of travel must be towards rather than static or away from the ideal.

There is no nation, no matter what political or economic system it employs, which exists today free of bribery and corruption.[29] But some are much better than others, and some are very much worse. No country where bribery is the basis of business-government interaction will ever achieve competitive advantage. Bribery and corruption work against all the Landes and Ferguson principles.

What is also clear from the discussion in this and the preceding chapter is that richness of natural resources is not what determines competitive advantage. South Korea has almost no natural resources and Japan is not particularly well-endowed either. But they have put in place the other Porter resources at a scale and quality far surpassing their global competitors. All this loops back to the first of the systems diagrams in Chapter 1, the discovery through research and development of a process to take a raw material and turn it into a manufactured product. A country can have all the proper political systems and capital markets, but only if there are ideas coming forward, ideas for processes that are better than what already exists, can a country achieve competitive advantage through materials, and materials are fundamental.

Advancement based on merit is important, and societies should not delude themselves into thinking that they have such a system when measurement shows that they are drifting away from it. Great progress has been made on removing race and gender as criteria for progression in many countries, but there is still a long way to go.

In the last chapter of this book there will be a chance to look to the future. Today, some economists and business strategists look at global movement of people, technology, trade and knowledge and think that national competitive advantage is no longer possible. By looking at the specific developments in materials for clothing, food packaging, information, buildings/infrastructure, electricity and transport in the next six chapters and how these were used historically to achieve competitive advantage, perspective will be gained about whether the Landes/Ferguson principles and Porter factors still apply in the future.

## Notes

1 Cited in the OECD publication *Defining Productivity*, 40526851.pdf (oecd.org)

2 David Landes, *The Wealth and Poverty of Nations*, Norton, 1998, p523

3 Philippe Aghion, Céline Antonin, Simon Bunel, *The Power of Creative Destruction*, Harvard University Press, Cambridge, MA, 2021, p179

4 Benedict Anderson, *Imagined Communities, Reflections on the Origins and Spread of Nationalism*, 2nd Ed., Verso, 2006. Anderson uses the term Imagined Communities because the people in the community/nation do not actually know one another, but still think of themselves as a community/nation and are willing to die on behalf of it.

5 Sewell Chan, 'Obituary for Benedict Anderson', *The New York Times*, 14 December 2015

6 David Landes, *The Wealth and Poverty of Nations*, Norton, 1998

7 There will be those who disagree with some of these principles of government, and especially the last one, arguing that there are things that government can do better than the market, and it is certainly possible to find examples of both smaller and larger government in competitively successful nations.

8 Niall Ferguson, *Civilization*, Penguin, 2011, pp305–6

9 Adam Smith, *Wealth of Nations*. I am indebted to David Landes for pointing me towards this quote.

10 Michael E. Porter, *The Competitive Advantage of Nations*, 2nd Ed. Macmillan, 1998. The section immediately following on factors is from pp74–75.

11 There has been a lot of criticism of Porter's analysis, though not particularly of these factors. For a review of the critical comments and a contrast between Porter's analysis of competitiveness and that of Paul Krugman see Nikolaos Alexandros Psofogiorgos and Theodore Metaxas, 'Porter vs Krugman: History, Analysis and Critique of Regional Competitiveness', https://mpra.ub.uni-muenchen.de/68151/1/MPRA_paper_68151.pdf accessed 4 March 2021.

12 It is certainly possible to see this description of the change in society through industrialisation as highly idealised and not reflective of what it was like for most of the workers. For a different viewpoint see Eric Hobsbawm, *Industry and Empire*, revised edition, Penguin, 1999. This book also presents a lot of data on wages and how they evolved during the Industrial Revolution.

Nonetheless, as far back as the 1720s and throughout the 18th century, Hobsbawm points out that French visitors to Britain noted that the 'whole British system was based, unlike that of … less prosperous countries, on a government concerned for the needs of … the honest middle class, that precious portion of nations… Britain struck the foreign visitor chiefly as a rich country, and one rich primarily because of its trade and enterprise; as a state of unusual liberty and tolerance – both of which were yet again closely allied with trade and the middle class. … it was unquestionably the most flourishing and progressive of economies, and one which more than held its own in science and literature, not to mention technology'. So while there was a lot wrong in Britain, there was less wrong than in France, hence competitive advantage.

13  D. Mate, A. Novotny, D. F. Meyer, 'The Impact of Sustainability Goals on Productivity Growth: The Moderating Effect of Global Warming', *Int. J. Environ. Res. Public Health*, 18, 11034 (2021), https://www.ncbi.nlm.nih.gov/pmc/articles/PMC8583465/

14  D. H. Meadows, D. L. Meadows, J. Randers, W. W. Behrens, *Limits to Growth*, Potomac Associates, 1972

15  'Roy J. Plunkett', Science History Institute, www.sciencehistory.org/education/scientific-biographies/roy-j-plunkett/#:~:text=Teflon%2C%20discovered%20by%20Roy%20J,and%20training%20to%20recognize%20novelty

16  An excellent popular book discussing invention with 35 specific examples and a lot of fun things to think about is Steven Caney, *Steven Caney's Invention Book,* Workman Publishing, 1985.

17  Michael E. Porter and Claas van der Linde, 'Green and Competitive, Ending the Stalemate', *Harvard Business Review*, September–October 1995. https://hbr.org/1995/09/green-and-competitive-ending-the-stalemate

18  Nathan Rosenberg (Ed), *The American System of Manufactures*, Edinburgh University Press, 1969, p366

19  An excellent discussion of this appears in Daniel Immerwahr, *How to Hide an Empire*, Vintage, London, 2020, who also makes clear how standards can be a source of competitive advantage.

20  For a particularly strong and articulate discussion of the importance of regulation of anti-competitive behaviour see Michael Porter, op. cit., pp662–9.

21 Harari, in his book *Sapiens*, op. cit., has discussed at length humans being able to create fictitious entities and practices, such as calling a company a person and double-entry bookkeeping, as characteristics that distinguish human societies from others.

22 Entrepreneurship & Business Statistics, GEDI, http://thegedi.org/global-entrepreneurship-and-development-index accessed 13 July 2022

23 Optimizing development processes for Smart Specialisation: Regional Entrepreneurship and Development Index as a tool for the design of regional Entrepreneurial Ecosystems | Global Entrepreneurship Development Institute (thegedi.org) accessed 13 July 2022

24 https://worldpopulationreview.com/country-rankings/gini-coefficient-by-country accessed 13 July 2022

25 Stefan Collini in *London Review of Books*, 1 April 2021, p15

26 See for example Daniel Markovits, *The Meritocracy Trap*, Allen Lane, 2020, and Peter Mandler, *The Crisis of the Meritocracy*, Oxford, 2020. These are discussed at length by Stefan Collini, op. cit. pp15–22.

27 'The Birth of the New American Aristocracy', *The Atlantic*, https://www.theatlantic.com/magazine/archive/2018/06/the-birth-of-a-new-american-aristocracy/559130/ 'The 9.9 Percent Is the New American Aristocracy' accessed 25 July 2022

28 T. Piketty, op. cit., Chapter 1.

29 The great anthropologist Emile Durkheim argued that crime and corruption were always going to be present in every society, although others have disputed his reasoning.

# 3. PACKAGING FOOD

*In each bottle, and at small expense, is a glorious sweetness that recalls the month of May in the depths of winter.*
Alexandre Grimod de la Reynière, author of *L'Almanach des gourmands*

*An army marches on its stomach.*
Famous saying of either Napoleon or Fredrick the Great

*PRESERVED FRESH PROVISIONS, BY DAGGETT & KENSETT, NEW-YORK; Warranted for any Voyage or Climate, In tin cases from 2lbs. to 8lbs. each. The prices here stated are for 4lb. cases of meat, and concentrated gravies, ready cooked, and without bone... The vegetable and gravy soups will be found cheaper, at the low prices here offered, than any nutritive and healthy fresh provisions can in any other way be furnished at sea. Plain directions for preparing these provisions for the table accompany each case... Concentrated Soups, in 2lb. cases, calculated to make, when diluted, a gallon of rich Soup; half gravy soups, half vegetable, $9 per dozen. Lobsters, Oysters, Clams, Fish, and the most delicate animal substances, for Sea Stores and Inland consumption, put up in order.*
1822 Advertisement in the *New York Evening Post*

## FOOD PACKAGING AND COMPETITIVE ADVANTAGE

The preservation of food through packaging materials – starting at the beginning of the 19th century with glass containers, moving quickly on to tin cans (steel-plated tin actually), and in the 20th century progressing to aluminium, Tetra-pak, and all sorts of plastic packaging – has had a profound effect on how people live. For the countries that had canning available before others, it provided a source of advantage to their navies and to all sorts of non-military long-distance shipping; for America, it meant that food could be taken on the long journeys to settle the west; ultimately for many countries, but particularly in the US and Western Europe, it made possible the growth of cities. In so doing it led to an increase in the productivity of agriculture. In the 20th century it changed the way people shop, replacing the corner grocery with the supermarket, and through the brand-driven printing on packaging influenced what people buy as well as how they buy it.

## WHAT MUST A FOOD PACKAGE DO?

Look into your refrigerator, the cabinets or larder where your food is stored. Viewed through the eyes of this book, what you see is an array of materials. Glass jars filled with jam, mustard and pickles; glass bottles of beer, wine, gin and vodka; aluminium cans of fizzy drinks; multilayer paper-based packaging for orange juice; plastic containers of milk and water; aluminium-coated plastic packets of crisps (potato chips); cardboard boxes of cereal, inside of which are plastic bags; eggs in specially shaped cardboard containers; perhaps fresh fruits and vegetables in some mix of plastic, cardboard and cellophane. True, this may look different if you are living on a farm, or are particularly conscious of reducing packaging, but as the world has urbanised food packaging has increased dramatically in quantity and variety. People have come to expect a variety of food to be available independent of season, and packaging is what makes this possible. The world uses about 180 billion aluminium cans every year. Polyester used in food bottling is a $250 billion industry. The Swedish-Swiss food packaging company Tetra Pak had revenues of ca. $15 billion last year.

These materials serve many functions, including:

- Containing the product, so the consumer knows exactly the amount they are purchasing.
- Protecting from contamination, preserving freshness, stopping damage from light or oxygen, or other airborne contaminants.

- Allowing for transportation, sometimes over long distances, without breakage or spoilage, while being as lightweight as possible to keep the cost of shipping low.
- Carrying information about the product and providing a vehicle for advertising or messaging.

Food packaging thus provides protection, utility and communications in physical, atmospheric and human environments. An optimum package provides all three functions in all three environments.[1] Food packaging is thus a major contributor to reducing food waste, while at the same time being a huge waste problem, as much of it is either not recyclable or discarded carelessly.

The archaeological record shows that from earliest times humans have developed materials to store food and drink. Clay, a silicate mineral, is very widespread as a raw material. When it is wet it has a plasticity that allows it to be shaped to form containers, following which high temperature firing makes the shape permanent. Pottery sherds and even intact vessels are found in Eurasia and North Africa, but some of the oldest, dated to 14000 BCE, have been found in Japan. It seems that humans almost everywhere learned to make pottery from whatever local clay they had. It was important for the maintenance of life, protection of supplies from animals, reduction of water loss through evaporation, as a defence against contamination – but it was not a source of competitive advantage because it was so easy to do. Likewise, cloth was very widespread (see Chapter 4) in antiquity, and undoubtedly was used to wrap up food as a form of protection.

To preserve food for longer periods, techniques such as drying, salting and pickling were developed. These techniques are also ancient, certainly going back beyond 2000 BCE. While they were all developed empirically, that is without an understanding of the science that makes them effective, they are useful techniques for killing bacteria and providing an environment in which bacteria cannot grow. The modern development of packaging for food preservation, from the early 19th century onwards, used this same idea, again without understanding the science until Pasteur's work, but in a realisation which was motivated by the desire to obtain competitive advantage.

## BOTTLING AND CANNING

Supplying food for an army far from its home base, or naval vessels out of port for weeks at a time, is one of the great logistical challenges of a military operation. Spurred by Napoleon's ambitions, even before he became emperor, the *Directoire*, as the government was called in late stages of the French Revolution (1795–1799), issued a challenge to develop a method for preserving a wide variety of foods, with a prize of 12,000 francs. It is not known how many people worked on this, as history only records the winner: Nicolas Appert. Appert was a confectioner who turned his attention to this problem in 1795 and worked on it for more than a decade. He showed his first results in 1806 at an exhibition, Exposition des Produits de l'Industrie Francaise, by this time having set up a business selling fruits, soups, jams and dairy products that he had heated and sealed with corks in glass bottles. The government declined to give him the prize, but he continued his work, expanding the range to include a wide variety of meats. In 1810 he was awarded the prize, on the condition that he make his method publicly available, which he did through publication of a book, *L'Art de Conserver les Substances Animales et Végétales* (The Art of Preserving Animal and Vegetable Substances), that same year.

Napoleon thus secured a method for obtaining more varied food supply for his army and navy, but, perhaps because of the spirit of the Revolution, France did not provide the intellectual property protection that allowed Appert or anyone else to grow an enterprise at scale. Despite many attempts at building the business, Appert died in poverty in 1841. Today he is honoured in France with many streets named after him and recognised as a pioneer in the world of food packaging science globally.

It was not just for military uses that preserving a variety of foods was needed in France. Indeed, one of the causes of the fall of the *Directoire* and the rise of Napoleon as emperor was severe food shortages in Paris. Sometimes securing competitive advantage for the military can improve the condition of residents at home, but more often it diverts resources from domestic priorities.

Appert's methodology did achieve commercialisation at scale, but it happened in Britain and America. Peter Durand was a British businessman on the lookout for opportunities. In 1810, just when Appert's work of 15 years became public, Durand was in France where he apparently learned of the development from a French inventor, Philippe de Girard. De Girard had the idea to shift from preserving food in bottles to metal cans. Appert had considered this but France did not have the quality of metallurgical expertise that existed in Britain at that time. The key technology was plating of tin onto iron, so that the (often acidic) liquids in the food products did not cause rusting. Although the details of the story are not recorded, it appears that de Girard discussed this with Durand and asked for help in patenting canning in Britain. Durand rushed home and did exactly that, except for the part about helping de Girard, receiving English patent no. 3372 in 1810 for preserving food in cans. For good measure he included claims for preservation in glass as well. So all the French ideas and developments that the French government insisted on being freely available became protected intellectual property in Britain.

In 1812 Durand sold his patent rights to two fellow Englishmen, Brian Donkin and John Hall, for £1,000. Donkin was one of the expert practitioners in the tinning of iron and saw the opportunity to expand into food. Donkin and Hall set up a canning factory and in just a year were supplying tinned food to the British Army and Navy. By 1818 the British Navy was using 40,000 pounds of tinned food on its ships.[2] But Durand was not finished. In 1818 he received a US patent for the very same invention. With our modern approach to patents this would not seem possible, but the patent system in the US was not very concerned with prior art at that stage, and one finds patents being granted for the same or closely related inventions that had already appeared in other countries. Thus, on 19 January 1825 Thomas Kensett and his father-in-law Ezra Daggett were awarded US patent 4009 for tinned food in the US.[3] Kensett, originally an engraver who had emigrated to America from England, was also patenting an idea he had seen in practice, but he worked for some years to make this into a proper industrial process. Daggett and Kensett became, for a time, a substantial company for the supply of tinned foods in America, for ships, cities, and for those exploring and settling the west. A testimonial to their product survives:

> *This is to certify that I took on board the Packet Ship Columbia, two cases of Fresh Beef and one case of Chickens, prepared by Daggett & Kensett, New York; they remained on board during our voyage to Liverpool, and back again to New-York; they were then opened in the presence of Capt. Browne and other gentlemen at the Fulton-street House and found to be fresh and good; and we have so favourable an opinion of Fresh Provisions put up as above, that we shall hereafter take them on board our ships. James Rogers Capt. of ship Columbia. Wm. Browne, late Capt. of ship James Cropper. New York Apr 17 1822. We ate of the Chicken and Beef taken to Liverpool and back to New-York and found them as sweet as any provisions we ate. [signed by 27 names]*[4]

The US, Britain and France were, in the early decades of the 19th century, the only countries with a variety of foods packaged and preserved

An early tin can, showing hole through which it was filled.

across seasons and geography. They exploited this for military advantage, both army and navy, but also to enable the growth of cities. In the US, can manufacturing prospered in the northern states along with other industrial enterprises such as clothing and shoe manufacturing. There was a great demand for tinned seafood, much of which was brought in at nearby New England ports. As previously mentioned, during the growth period of western settlement, wagon trains were stocked with tinned food, especially vegetables. While Kensett's original business ultimately failed, the place of canned food in the American diet was secured. Thomas Kensett died of consumption, aged 43, when his son, also Thomas Kensett, was 15. The younger Thomas proved himself an astute businessman, and after building up a successful cloth business, resumed the canning business in Baltimore in 1851. This coincided with a second wave of westward expansion, driven by the California Gold Rush of 1848, and he found a growth market for his goods with those heading west.

Competitive advantage to the military from tinned food appeared again with the start of the Civil War in 1861. Northern troops were fighting far from home. While when they were successful in battle they could loot conquered farms for food, their day-to-day needs were met by the supply of canned food.

Cans from the early 1800s did not look like those of the 20th century. Iron with tin plating was shaped into a cylinder, then a top and bottom were formed as caps that could be soldered around the sides to provide a seal. A hole was cut in the top to allow filling of the food. After the heating process was complete, a metal disk was soldered into place to seal the container.[5] Lead in the solder probably caused occasional poisoning of the consumer.

Comparing the cans of the first half of the 19th century with all later ones, one big difference was the lack of a rim protruding from the top. This had a function in sealing, to be sure, but it also made it possible for a can opener to hook on and leverage to be applied to puncture the top. This was not needed in the early cans – because the can opener had not yet been invented! Cans were opened with a hammer and chisel – a soldier sometimes used a sword or bayonet to puncture the lid. While throughout this book there are many examples of key inventions taking place at about the same time or even before they are needed (the pencil and fountain pen more or less coincident with mass production of paper, the Otis elevator brake before the steel construction technology for the skyscraper) it is inevitable that sometimes this doesn't happen. The first can opener was of a type that was still around when many readers of this book were growing up; it was patented by E. J. Warner in 1858 and used a sharp knife to puncture the can, the long handle providing the necessary lever arm, and the remainder of the mechanism to hold the opener in place. These became widely available, cans were manufactured to suit the opener, and the openers were an essential piece of the kit of Union soldiers in the Civil War. The can opener with a cutting wheel, which was in one form or another the standard until aluminium cans were provided with pull tabs to open, was patented in 1870 by William Lyman.

The patent system, in both Britain and America, even with its afore-mentioned flaws, played an important role in spurring food packaging innovation during the 19th century. One more invention at the end of the century has survived in the market to this day, as has the company that was created around it: the bottle cap, invented in 1892 by William Painter.

The inside of the cap, now plastic, was originally cork, and it made possible the tight sealing yet easy opening of glass beverage containers, especially carbonated beverages. Painter founded the Crown Cork and Seal Company in Baltimore, which remains a major force in the food packaging industry, today known as Crown Holdings. Crown made many further innovations in food packaging technology. Its culture of invention led one of its employees, King Gillette, to invent the disposable razor and start his own company, having seen the business opportunity in a bottle cap as something that is used once and then thrown away.

## BEYOND CANS

Paper, cloth and cardboard certainly predated the use of cans for food packaging, with paper and cloth being especially suited to cases where flexible packaging was required. This was widespread and, like early ceramic vessels, not a source of competitive advantage. Anyone who had cloth could use it to wrap up some food for protection or transportation. Paper was less useful, because it was too valuable when the main source of fibre was rags, before the days of wood pulp (see Chapter 5 where this is discussed in more detail). Still, flexible packaging was desirable and destined from the last quarter of the 19th century onward to play a major role. Once paper became cheap, British customers got their fish and chips wrapped in yesterday's newspaper.

The development of wood pulp into paper yielded many grades with varying properties. The finest paper, for writing and printing of books, was white, but less intensive processing (less bleach) gave brown paper, and the machinery could allow this to be thicker and stronger. The Kraft process, invented in the late 1870s by Carl Dahl, and first commercialised in Sweden, opened up cardboard as a major packaging material, of which food was a substantial part of the market. (For a discussion of the chemistry of paper from wood pulp see Tech Talk 4.)

During the same period (1850–1900) machinery for making paper bags was developed. The initial use was for large bags to replace cloth sacks for flour and other bulk food products. Indeed, a view of the food production cycle for processed foods such as flour or even rice is that it is a sequence of breaking bulk. One starts with very large quantities stored in bins or tanks, which are then put into barrels or sacks. Historically these were of sizes using multiples of the English unit of weight the stone, which is 14 pounds, hence flour sacks of 140 or 196 pounds. Before materials for packaging became cheap and widely available, a grocer would take delivery of such a sack and sell smaller quantities from that. This was economical in terms of packaging materials but not in terms of waste and contamination.[6]

After some resistance on the part of railroads and other shippers, corrugated cardboard replaced wood as the dominant packaging for many food products in the late 19th/early 20th century. This was followed by smaller cardboard boxes taking an increasing role in packaging for the final consumer. Breakfast cereal was among the first food products to take advantage of the inexpensive cardboard box of the late 19th century, with Kellogg cereals being sold in boxes starting in 1906.

Machinery to make a flat bottom glued paper bag, similar to bags which are still used today, was invented in 1868 by Margaret Knight, a prolific inventor (her first major invention for enhanced safety in cotton mills was made when she was only 13) and a prolific filer of patents. She built a scale model of the machine, patented it, successfully defended her rights against a man who attempted to steal her invention, founded the Eastern Paper Bag Company, and developed a system to licence and receive royalties.

Even in the early days of paper and cardboard packaging, it was clear that the package did not just contain the food but could be printed to provide information about it. Paper labels glued on to cans or bottles served the same function. Cereal scooped out of a large sack at the grocery into the customer's container is just cereal, but in a printed box it is Kellogg's Corn Flakes. Beans in a can become Heinz Baked Beans, 57 Varieties. Over time cereal boxes printed stories or pictures of film characters to appeal to children, ingredients, and, in more recent decades, nutritional information.

The printing on the box also provided a means of protection and assurance to the customer that the contents were genuine and not a copycat product. It was only in the last third of the 19th century that the trademark system was developed in the US, Germany, France and the UK as part of the protection of that aspect of intellectual property. The first food-related use of this was probably by the Smith Brothers for the protection of their cough drops in the US against competitors, using the words Trade and Mark on the package to prove to the customer that it was their special formula inside.

# Kellogg's

A distinctive script, such as that developed by Kellogg at the very beginning of the company, both helped to distinguish the product from competitors, and protect it from copiers.

**TRADE** **MARK**

# SMITH BROS.

## TRUSTED SINCE 1847

Both images use the distinctive letter R in a circle, ® indicating a registered trademark.

In Europe before World War I most grocery shopping was done by women who were served by the local grocer. He made use of the developments in paper bags to repackage material he had bought in bulk. But in the US this had already begun to change, with the first self-service grocery stores, the forerunner of supermarkets, opening in 1916. Ruben Rausing, a young Swedish graduate of the Stockholm School of Economics, went to the US to study for a graduate degree at Columbia University. He saw the attraction of the self-service grocery and that this would increase the demand for pre-packaged goods. In 1929 he co-founded a company specialising in food packaging, Åkerlund & Rausing, that was to become one of the largest European food packaging manufacturers by making various kinds of cardboard boxes for dry foods. While much of this dry food began to be sold as self-service, perishables did not. This was particularly true of milk, which because of limited shelf life and requirement for refrigeration continued to be delivered to homes several times a week. Rausing worked to develop a way to package milk for long shelf life, building on everything that had been learned in the previous century about the need for aseptic packaging. However, he made a radical change to the process. In canning, the food and the container are combined and then rendered aseptic; Rausing pursued an approach whereby the milk and the container were sterilised separately and then combined.

Machines for folding and cutting paper had been in use, for example for bags, but Rausing devised the tetrahedral package using heavy paper with a polyethylene coating. The coating technology was itself innovative. The package could be filled with milk and then sealed aseptically below the level of the liquid, ensuring that there was no air space to cause contamination. This new package met all the criteria for food packaging – it was light and easy to transport, protected from the environment, and the amount in the package was known with precision. It was also inexpensive and had very little waste in any of the materials, so that it could compete from the outset with other dairy packaging.

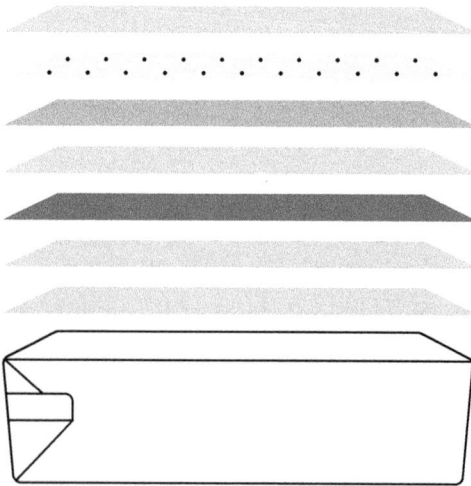

The layers of the Tetra Brik Aseptic package, from the outer layer inwards: polyethylene; printed design; paper; polyethylene; aluminium foil; polyethylene; polyethylene.

Working from the company facilities in Lund, Sweden, it took from 1943 to 1951 for the machinery and sealing technology to be ready for commercial exploitation.[7] Today's business executives who want the research and development process for materials innovations to go from idea to commercial realisation very quickly would do well to learn the lessons from this and other high-impact developments. The first cream was packaged in tetrahedra (now known as Tetra Classic) in 1952 at a small dairy in Lund. Seven years later, in 1959, Rausing's new company, Tetra Pak, produced 1 billion containers! But R&D did not stop. The company developed the brick (Tetra Brik) in 1960, and the gable (Tetra Rex) in 1965. In 1971 Tetra Pak sold 10 billion containers, and by 1977 20 billion. This doubled again by 1986, always driven forward by a stream of innovative developments and global expansion. As everyone who shops in supermarkets today knows, it is not just dairy products, but juice, soup, wine and much more that is in a Tetra Pak package. Today's package is a multilayer container, embodying a range of materials put together to achieve sophisticated requirements.

Ruben Rausing got his first patent in 1944, and Tetra Pak has remained an active user of the patent system for its packaging technology. But this company teaches some other important lessons about competitive advantage from novel materials technology. If one reads Michael Porter's classic text, *The Competitive Advantage of Nations*, it would be natural to draw the conclusion that a country like Sweden could never produce companies that innovate, grow to scale, make its founders wealthy,[8] and achieve a world-leading position in its industry. While Sweden certainly has eminent research universities/institutes, a highly educated population, and strong rule of law, it also supports strong socialist principles in welfare, health care, high taxation, coupled with a general suspicion of such tools as stock options and uncontrolled free markets. Yet it is not just Tetra Pak, but a plethora of important global companies such as H & M, Ericsson, and IKEA that have originated there. Sweden teaches that it is too easy to believe that a single model of capitalism, essentially a US model, is the only one that can be successful. Sweden is one of several Northern European countries making different choices based on a set of values, while still being successful in business on a global stage. The growth of enterprise, and its concomitant national competitive advantage, requires certain principles (as described in Chapter 2) but does not demand a particular model for the State.

Tetra Pak also teaches that society has agreed that competitive advantage must be achieved within certain rules. Once a company achieves a certain market dominance, how it maintains that dominance must accord with principles of fairness. Tetra Pak made acquisitions and mergers, most notably with Alfa Laval but also several smaller companies, acquisitions that faced considerable scrutiny from regulators as Tetra Pak sought to vertically integrate and remove some smaller competitors. In the end it usually prevailed, but the level of scrutiny was appropriate to its position in the market.

More seriously, it was charged by the European Union with abuse of its dominant position in a landmark case in 1991, because of which the company was fined 75 million ECU. In the decision, it was found that the company had restrictive contracts designed to prevent competition from competitors. Tetra Pak obliged customers, mainly dairies, to stay loyal to its products, at the expense of actual or potential competitors, through the use of restrictive clauses in its contracts. These included an obligation on users of Tetra Pak packaging machines to use only cartons made by Tetra Pak and supplied under its direct control. This enabled Tetra Pak to assure customer loyalty artificially, thereby excluding carton competitors and securing revenue on the sale of cartons for as long as each machine was in operation. In all cases, Tetra Pak made its product guarantees dependent on this commitment. This, together with other aspects of its policy (for instance the fact that distribution of its products to dairies and other customers in the EU was performed exclusively by companies within the Tetra Pak group), had the effect of prohibiting customers from using either competing brands or Tetra Pak's own brands acquired on potentially more competitive terms.[9] Tetra Pak also sold some of its products at a loss to stop competitors from gaining a foothold, using the profits from others to subsidise the loss. In its most extreme practice, it even bought the total supply of a competitor's machines to keep them from being sold to customers.

Competitive advantage is a good thing, and this book both analyses and celebrates its achievement. But society is right to provide constraints that ensure that the advantage is not used to prevent others from competing in the same space. Centuries earlier, countries forbade those with knowledge of how to make something from emigrating or otherwise passing on the knowledge to others. To some extent such clauses are still allowed in employment contracts, though without the death penalty that older rulers imposed. As the Tetra Pak case and others have shown, while dominant positions can be achieved they cannot be abused. Strong regulatory regimes in many parts of the world have been put in place to define what constitutes abuse, and what the sanctions are for violation.

## COMPETITIVE ADVANTAGE TO NATIONS FROM CANNING, PAPER AND CARDBOARD

A theme running through all these developments in food packaging is that innovative, large scale, globally significant businesses depend on a foundation of the protection of intellectual property. Countries with a system where patents could be filed, examined, granted and defended in court became centres for food packaging innovation. This protection covers materials (as defined in Chapter 1), methods of sealing or opening a package, machinery, and sometimes the contents. Trademarks built on this protection, as well as offering assurance to the consumer about the provenance of the product they were buying.

The ability to preserve a wide variety of foods provided significant military advantage in the early 19th century to Britain and France, especially for their navies, but also for armies, and this continued with the Union Army in the US Civil War. Of course it was not only for the military but for crews and passengers of all long-distance shipping. The various advertisements by Daggett and Kensett show that canned foods facilitated the westward expansion of the United States, territorial acquisition that provided the foundation for a long period of competitive advantage.

All of this was important, because a source of sustained competitive advantage provides wealth that spreads through the economy. But it was not transformational. Food packaging did alter human society in such a transformational way, however, through facilitating urbanisation. Although data vary in early years, in 1800 England and Wales had at most one-third of its population living in cities.[10] This is not to say that the remainder were agricultural, far from it. But many lived in so-called market towns, supplied with food from the surrounding agricultural regions. By 1900 the urban population exceeded 75 per cent. Much of this urbanisation is due to the growth of northern cities such as Manchester, Liverpool and Leeds, driven by the Industrial Revolution as discussed in Chapter 4. But London also powered forward as a city during the 19th century, its population expanding from 1 million to 6.7 million.

This could only have been possible with the concomitant growth of food packaging. As cities grow, they inevitably reduce the amount of nearby agricultural land, and their supply regions become increasingly distant. Even when the distance is not so great in miles, it is in terms of time – how long it takes to get food from farm to store. Until the rampant growth of urban areas and suburbs between New York and Philadelphia from the 1950s onwards, southern New Jersey was a fertile area for growing vegetables, and produced a large tomato crop. To be sure, some of this found its way directly to urban tables through markets, but without canning most of it would have been wasted. Campbell Soup Company, founded in this region in 1869, produced canned tomatoes, soups using the tomato base, and many other products, gradually expanding to be one of the largest food processing and packaging companies in the world. Campbell and companies like it made the growth of cities possible.

In a country where cities can grow and thrive and develop dynamic economies that innovate and respond to changing demand, a sustainable source of wealth and competitive advantage can be achieved. Jane Jacobs discussed this eloquently in her book *Cities and the Wealth of Nations*.[11] While her thesis is more about the role of cities in manufacturing rather than services, the importance of cities that she describes was true in the 19th and 20th centuries and remains so. As she says, cities provide 'enlarged markets for new and different imports consisting largely of rural goods and of innovations being produced in other cities; ... increased numbers and kinds of jobs; ... increased transplants of city work into non-urban locations as older enterprises are crowded out; new uses for technology, particularly to increase rural production and productivity; and growth of city capital'.

It is not just the wealth and consumption of cities that leads to competitive advantage. As workers are drawn from rural locations to cities in search of better opportunities, farming must become more productive both in terms of yields to meet increased demand, and less labour intensive because the labour is just not available or must be imported from other countries. Certainly there is an argument to be had as to cause and effect, whether the growth of cities required improved agricultural productivity, or reduced labour requirements of agriculture led to migration to cities, but the result is the same. The US was particularly well organised in this drive for improved agricultural productivity to secure competitive advantage, through its system of Land Grant universities and the Agricultural Extension Service, developing more productive products and processes, then teaching farmers how to implement them. In this way government played a role in encouraging and developing efficient enterprise. These increases in productivity throughout the economy proved a significant source of competitive advantage.

## FOOD PACKAGING TIMELINE

| | | YEAR BP |
|---|---|---|
| 1795 | French government offers prize for food preservation | 227 |
| 1810 | Prize won by Nicholas Appert | 212 |
| 1810 | Durand gets UK patent for canning | 212 |
| 1825 | Thomas Kensett patent for tin cans in America | 197 |

| | | YEAR BP |
|---|---|---|
| **1850** | Introduction of paper sacks for flour | 172 |
| **1858** | First can opener invented by E. J. Warner | 164 |
| **1858** | Margaret Knight invents glued paper bag | 164 |
| **1870** | Rotating wheel can opener | 152 |
| **1906** | Kellogg's cereal sold in boxes | 116 |
| **1922** | Invention of aerosol can (but first commercialised two decades later) | 100 |
| **1929** | Akerlund and Rausing founded in Sweden, company which became Tetra Pak | 93 |
| **1935** | First sale of beer in a can | 87 |
| **1950** | Biaxially oriented polyester – Mylar | 72 |
| **1952** | Dairy product sold in tetrahedral package by Tetra Pak | 70 |
| **1957** | Beverage cans made of aluminium | 65 |
| **1962** | Pull tab can | 60 |

## Notes

1 Sara J. Risch, *Food Packaging History and Innovations, J. Agric. Food Chem.*, 2009, 57, pp8089-8092

2 Donkin and Hall's company survived and prospered, and in 1839 was acquired by Crosse and Blackwell, an old British prepared food company who wanted the expertise in canning. The Crosse and Blackwell name can still be seen in specialised food shops today.

3 19 January is commemorated as Tin Can Day in America in honour of this patent awarded for someone else's invention 15 years earlier!

4 https://www.archives.gov/files/publications/prologue/2013/fall-winter/patents.pdf accessed 23 June 2021

5 For a video demonstrating how an original tin can was made see https://www.bbc.co.uk/news/av/magazine-22093655 accessed 15 June 2021

6 It is interesting that during the Covid pandemic in 2020, when there was a sharp increase in demand for flour and yeast for home baking, the shortage of supply in supermarkets was not due to lack of flour and yeast but of packaging material.

7 See https://www.tetrapak.com/en-gb/about-tetra-pak/the-company/history for a history of the Tetra Pak series of innovations.

8 In the 2019 Sunday Times Rich List the Rausing family wealth was estimated at £9.6 billion.

9 https://ec.europa.eu/commission/presscorner/detail/en/IP_91_715 accessed 23 June 2021

10 Urban Population Data 1801–1911 http://doc.ukdataservice.ac.uk/doc/7154/mrdoc/pdf/guide.pdf accessed 15 June 2021

11 Jane Jacobs, *Cities and the Wealth of Nations*, Random House, 1984

# 4. CLOTHING

*Whoever says Industrial Revolution says cotton.*
Eric Hobsbawm, *Industry and Empire*[1]

In September 1991 Helmut and Erika Simon were walking off a path in the Tyrol, just along the Austria-Italy border. They were startled to see a body partially buried in ice, and alerted the authorities, who assumed it was a deceased mountaineer. They were right, but his was not a recent death. This was the body of what became known as Ötzi the Iceman, who had died around 3300 BCE.

Much has been studied about Ötzi, but one of the most remarkable things about this find is that a full set of clothing was recovered, including hide coat, skin leggings, fur hat and hay-stuffed shoes. The hide coat was mostly made from sheepskin, with some goat leather. The leggings were goat leather, the hat was bear fur, and he had a sheepskin loincloth stuffed with grass matting.[2] The shoes themselves were quite complex, using bearskin for the soles, deerskin for the upper panels, some tree bark netting, and laces made from cattle leather. All these items and much more can be seen in the South Tyrol Museum of Archaeology.

Quite a few animals must have been killed to provide clothing (and presumably some food) for Ötzi during his 45-year life. They needed to be of several species because different skins are more suited to one function than another. Some simple tools were required to cut and stitch the skins together, and it is possible that the fashioning of these tools, such as bone needles, gave *Homo sapiens* an advantage over Neanderthals. Ötzi's clothing did more than just keep him warm in the Alps. It also had a protective function, whether from sharp objects penetrating the soles of his feet or biting insects/reptiles attacking his legs. It was not armour, however, so it did not protect him from the enemy who eventually killed him.

As humans spread out from the lush African habitat where they originated, into colder, harsher climates, clothing became a primary need that became increasingly difficult to fulfil. It is not just the availability of skins that is the problem, but their properties as well. Clothing needs to allow a flow of air and moisture in both directions, a property that works well with most kinds of cloth but poorly with the skin of a dead animal. In effect, an animal's skin serves as a distinctly different kind of barrier for the animal, just as our skin does for us, from the barrier we want as our clothing.

Even in warm climates, and not just for protection, humans have an instinctive need to cover themselves, at least to use some form of clothing to cover their genital regions. Anthropologists have researched the possible reasons for this – avoiding distraction, protecting sensitive areas, showing who belonged to whom – but for whatever reason there has long been widespread demand for clothing, however minimal.

A nation of people living in a temperate climate needs warm clothing to survive and thrive in the winter months, and different clothing for the summer. It may need particularly durable clothing if it is largely agrarian, and completely different clothing for most urban living. Egyptian linen and Chinese silk met domestic needs, but, as a consequence of 'national' attitudes to export, didn't lead to competitive advantage. By contrast, the development of diverse grades of wool and cotton as clothing fabrics, and the mechanisation of production in Britain certainly did. Using both domestic (in the case of wool) and imported (from the US and India primarily in the case of cotton) materials, Britain produced clothing products far in excess of internal needs, generating great wealth for the country. It happened in a national structure that rewarded innovation and industrialisation. Using this wealth, in part derived from exploitation of its empire, Britain was able to expand and defend itself as a major power on the world stage, achieving substantial advantage over its traditional rival France.

In the 20th century, the development of synthetic materials for clothing led to the democratisation of fashion, and massive change in how the world was clothed. This was led by technical developments that occurred primarily in the United States, Britain and Germany, building on the world-leading academic chemistry establishments of these nations. One way to view the cheap materials that resulted is as an increase in productivity. Moreover, Nylon, one of the first commercial synthetic fibres, played a significant role in World War II for the US and Britain, and thus was a source of competitive advantage for military operations.

The story begins with linen.

## LINEN

Humans, and (as was learned recently) even Neanderthals, learned how to make string from plants, and later from the furry or woolly coats of animals, a very long time ago, and weave it into cloth. Archaeological evidence shows that this ability was very widespread, so just knowing how to make thread and weave it into cloth was not in itself a source of advantage. (It has been speculated that having string was a source of competitive advantage, because of the ability to wrap up a package and carry it.[3]) The plant that emerged early as being favoured for this was flax, from which linen is woven. Other, more ancient sources of fibre were hemp, jute, banana and other fibrous plants, depending on the climatic conditions suited to their growth. All of these are known as bast fibres, and they are appealing to the fabric producer because the individual fibres are very long. In Egypt the history of linen cloth using bast fibres derived from flax stretches back 7,000 years. The greatest samples of it, of course, are from wrappings of mummies, from which it is clear that large cloths could be made, with very fine weave, and in great quantity. Egyptian linen was also used extensively for sails.

Linen has all the attributes one wants in a garment that are lacking in skins. It feels cool on the skin in hot climates, it is strong so does not tear easily, and it holds up well with age, even becoming softer after a few years. Flax is well suited to the Egyptian climate, and they developed both the cultivation of the raw material along with the spinning and weaving processes. What Egypt did not do, over the millennia of domination of linen manufacture, was use this as a means of competitive advantage. True, there was trade in linen along the Mediterranean, but this was mainly done by Phoenician traders who grew wealthy from it as individuals or families but did not enrich any nation state.

## SILK

The other ancient source of fabric was the silkworm. There is evidence of silk being made in what is now China more than 8,000 years ago. And for at least 6,000 years the Chinese were the only source of silk in the world.[4] Now, that is a long-standing monopoly! For most of this period, i.e., until about 2,000 years ago, very little silk was exported. Indeed, export of the silkworms was a crime punishable by execution. After that, a lively trade in silk developed from east to west, and soon the cultivation of the silkworm on mulberry tree leaves (sericulture) was established in Persia, with attempts (usually failed) in other places as well. The Japanese learned silk production from China and Korea in the 3rd century CE, and it became their dominant product for export, accounting for about a third of all Japanese exports as recently as 1930.

One of the important characteristics that emerged with silk (and was in part true for linen as well) is the association of fine examples of the fabric with wealth and prestige. Silk appeared in Egypt and there is evidence of it being worn by Cleopatra. As with linen, it has often been found in the tombs of the wealthy. As it spread westward one finds this association of silk with the upper classes in the Roman Empire, Scandinavia, Russia and later in England. Other fabrics, particularly coarser forms of cotton, were worn by lower classes; denim became known as tough clothing for farmers and outdoor workers. An important theme of the materials involved in clothing is that what a person wears is an indicator of their societal class. This was one driver of the policies and practices of nations as diverse materials became available.

As with linen in Egypt, the Chinese did little to use their dominance of silk to achieve competitive advantage. There was some use to persuade marauders from outside the Great Wall to stay where they were by bribing them with silk, so maintaining a fragile peace. But the trade in silk along the Silk Route from China through Central Asia mostly enriched the traders, who did take enormous risks travelling great distances.

Another aspect of silk and linen that contrasts with materials to be discussed shortly is that the important research and inventions necessary to be successful are about the production of the raw material, rather than conversion of that material into manufactured product. The key thing the Chinese did well was to develop sericulture to high productivity, so they had large amounts of thread available. Similarly, the Egyptians worked hard at flax cultivation. Yes, there were progressive innovations in looms, but these were incremental advances rather than any sort of major advance that would lead to sustainable advantage.

With silk there emerges one other characteristic of cloth production that was replicated elsewhere over centuries: the notion that spinning and weaving were women's work. The implications of this for societal structure around the production of these materials will be discussed further when wool and cotton are considered.

So linen and silk are ancient. In the case of linen, its production was fairly widespread. By contrast, for silk the Chinese had a monopoly on the raw material that lasted for millennia. But in neither case do these fabrics provide a convincing case for competitive advantage exploited through the material.

All this began to change with wool.

## WOOL

The domestication of sheep is thought to have occurred in Mesopotamia more than 10,000 years ago, and they were among the first animals to be domesticated. Archaeological evidence appears to show that sheep were used as a supply of meat and milk, while their woolly skins were tanned and provided warm clothing. About 8,000 years ago in what is now Iran there began the selection of sheep for wool, but it would be another two to three thousand years before the first wool garments appeared. These may have been felt (a mat of woollen fibres made by pressing together the wool, usually with water and heat) rather than woven wool. The expert weavers of the Middle East concentrated on linen from flax because the fibres were longer than those in wool, and warmth was not such a consideration for them.

In Chapter 1 it was asserted that a nation cannot obtain competitive advantage through raw materials; usually this just leads to exploitation. This is true even for the fibre required for clothing. Indeed, it remained true for the cotton growers of the southern US and India for many years. But there is a subtle distinction to be made in the case of some living species that produce these fibres, which was true for silkworms (but not exploited) and comes to the fore with wool.

Before anything was known about the science of genetics, farmers knew about selection for certain traits. Darwin was influenced by agricultural selection processes – the farmers didn't know about Darwin's theories, but Darwin knew about the farmers! Over a period of 300–400 years, and in several countries, farmers selected sheep for quantity and quality of wool. Quality relates to properties of softness, fineness of fibre, length of individual fibres, and other properties that enhance the ultimate fabric. Quantity is clear: the amount of wool produced per sheep. Rams with desirable quality and quantity of wool were highly prized and could be moved over long distances for breeding.

So when sheep owners select and enhance their breeds, they are increasing the value of the wool, and if (as was the case in the Middle Ages for both Spanish and English breeders) they do not freely export the sheep, there is the possibility of establishing a sustainable competitive advantage based on a very desirable (raw) and continuously improving material. This was true for both England and Spain in varying degrees between approximately 1200 and 1600. The Spanish developed what became known as Merino wool, probably starting with sheep brought to Spain from North Africa. Selective breeding to enhance properties is not like an invention that revolutionises an industry. It requires the long view, i.e., it is about slow improvement over centuries, especially with an animal like sheep that only gives birth once a year (silkworms are a better bet in this regard). This may be the longest running research and development project in human history. The improvements they achieved were remarkable. Just looking at quantity of wool, the Merino sheep grow their coats continuously, so they only survive if they are sheared regularly. They are living wool-making factories, only requiring grass and water as input material.

English and Scottish sheep in the Middle Ages were also the result of extensive breed management, made possible by the economic and social system around the large manor house farms. There were three main breeds: the Ryeland, which were short-wooled, as well as Cotswold and Lincoln, long-wooled sheep that dominated the industry for many centuries before the rise of the Merino.[5] All of these produced a very fine wool, and hence a soft cloth, contrasting with the wool of other sheep in Europe.

For a long time, the English were content to export their wool, mainly to European centres such as Florence and Belgium/Holland for cloth production. The supply of this material to the weaving centres was so crucial to the economies of the importers that if it were interrupted (for example by war, weather, or disease) mass starvation could occur, as well as social disruption when people were required to migrate to find work. The English kings understood the power this gave them. They also recognised that taxation of the wool trade would provide a source of wealth to the Crown, and they used this to fund their military adventures. Thus in 1297 Edward I used taxation of wool to fund his army in the invasion of Flanders. This unsuccessful attempt at conquest interrupted the flow of wool, and though the Flemish triumphed over the English army they suffered great economic damage. One estimate is that in 1297 wool constituted half of the English economy! A wealthy wool merchant of the time had engraved on the window of his house:

*I praise God and ever shall*
*It is the sheep hath paid for all.*

Fine wool, as an advanced material of the 13th and 14th centuries, became central to the way the king and Parliament established and utilised their competitive advantage. The most basic approach was to tax all exports of wool. This was done with the agreement of the merchants who traded the wool to continental Europe. But the kings went beyond this, taxing the wool itself at source (generally the great manors in areas like Lincolnshire, who often had flocks involving several nearby manors cooperating in all aspects of production, and also at large monasteries) with the

Crown taking payment in wool. Kings also sometimes took loans in wool, which they would eventually repay as expected revenue was received. All of this financed both the ordinary expenses of the king and court, as well as extraordinary expenses such as invasions and defence.

In such a system, it takes either wisdom or a government with checks and balances to keep it from going out of control. England was only beginning to develop the latter by the time of Edward III (reigning from 1327–1377), with Parliament's role in taxation in a developmental phase. Tax on wool exports had been in place since the reign of Edward I (1272–1307), but with the start of the Hundred Years War in 1337 the needs of the king became ever greater, and there was not a sufficient brake on the various forms of taxation of wool. This very nearly ruined the industry, and certainly dissipated some of the huge competitive advantage that had been secured over centuries. Still, the last chapter had not been written.

The supply chain disruption caused by the Hundred Years War greatly affected the ability of the Flemish weavers to practice their trade. Because of the war England also suffered from difficulty in exporting its wool, which in previous times attracted very high prices. The parties (that is, the wool producers and the weavers) had a mutually advantageous solution: relocate the Flemish weavers to England. Weaving is a skill, but Flanders had not managed to make it a source of national competitive advantage. Now in England the wool production and the cloth production were, over several decades, integrated into a formidable industry, because of which England greatly reduced its export of wool and increased its export of cloth.

Still, it was necessary to take the production of woollen garments to a different level if true competitive advantage was to be achieved. The Flemish weavers, as well as their Welsh counterparts (who often worked with lower quality wool) were almost entirely single cottages, small family businesses. This was, in a sense, fine, when there was no mechanisation of the production. But as soon as machines were able to take over some of the process of producing woollen cloth and garments, there was an opportunity to move into larger industrial production. The first big centre for this was Leeds. In the late 1400s Leeds had a population of about 1,000. This tripled in the next century, and then doubled again so that in 1650 it was about 6,000. By 1801 it reached 94,000 and by 1851 it was 250,000. Throughout this period, the dominant industry was production of woollen fabrics. Around the wool industry grew major engineering capability and foundries, useful for making machinery, but then able to provide support for other industries. Financial services also were developed, so that even now, when many of the manufacturing industries have declined, the city remains an important financial centre.

Crucial to the development of this inland town/city as an export centre was infrastructure, in particular the Leeds and Liverpool canal that opened in 1816, followed by extensive railway links in the 1830s. Neighbouring Bradford, also becoming a major industrial centre, benefitted from this infrastructure as well. The canal carried coal and limestone from Yorkshire to Liverpool, and when it was first conceived in the last quarter of the 18th century that was the main motivation. By the time it was opened 50 years later, revenues from merchandise such as textiles heading from Leeds and Bradford to Liverpool for export equalled those from coal and limestone. Building the canals (for there were several) and railways required a cooperation and shared vision by government and private financers. This joint effort on the part of government and capital providers in support of the critical role of enabling infrastructure is a feature of nations that achieve competitive advantage associated with materials.

Common to both the mechanisation/industrialisation of the wool industry and the construction of the infrastructure necessary to support the trade was engineering.[6] The first steps in mechanisation only employed the most fundamental machines, but they automated some of the most odious jobs. For example, woollen cloth undergoes a process called fulling. This is a pounding of the cloth, immersed in liquid, which removes rough fibres that are protruding, while washing away unwanted fats and oils. Fulling has been called one of the worst jobs that ever existed. In Roman times slaves did it, and in more recent centuries in Scotland it was done by women. The textiles were immersed in a bath of human urine and the fullers had to pound them with their feet. Early in the mechanisation of wool production, before the Industrial Revolution was even gathering momentum, English engineers used water wheels as a source of energy to drive a hammer in the fulling process, and even designed in a rotation of the cloth so that it was treated evenly. The deep competency in engineering for invention, coupled with rewards, both financial and reputational, was to play a crucial part in the huge competitive advantage that Britain obtained from its textile industries, starting with the early mechanisation of steps in woollen textile production.

When a country such as Britain is driven to secure maximum competitive advantage from an industry such as wool, it does not stop when all of its domestically produced wool is being used by its newly constructed mills. Going beyond domestic needs was made clear as a route to advantage in Chapter 1. Other countries, such as Australia, became valuable as places in which production of the raw material could be increased. These suppliers needed to be part of a prosperous market controlled from the centre, lest they start to offer their wool to the highest bidder, or seek other alliances based on their wool. Hence, the wool industry became important to the case for empire. In this configuration, the competitive advantage of the breeders could remain with the controlling country, as would the value addition of the mills.

Likewise, there is no reason to be satisfied with the domestic market and nearby (European) markets for the products of the mills.[7] While the nearby markets might have wanted the most sophisticated (in terms of style, softness, colours), the slave owners in the American South and the Caribbean liked plain wool clothing for their slaves, wanting it to be cheap and durable. English and Welsh woollens, undyed, sold for this purpose were known as kersey and white Welsh plains, and mixtures of wool with cotton was called homespun, while there was a linen/wool blend called linsey-woolsey.[8]

The sustainable competitive advantage that Britain achieved in wool and woollen textiles spanned more than six centuries. It certainly evolved over that time, but throughout was characterised by strong coupling between private enterprise of the sheep breeders/wool producers with the mill owners and exporters. It drew on a broad range of skills in the population, from animal husbandry to engineering. From the 13th century onwards it both financed the military and was enhanced by the success of conquest.

It helped provide the wealth that could finance infrastructure, then benefitted from the canals and railways. If one looks at any brief period of this successful industry, there are good and bad times, military defeats, plagues, bad government, successes and failures at every level. Nonetheless, in overview, looking back to this extended period of 1200–1900 wool represents a critical and remarkable success for Britain.

But it was only the beginning. Because halfway through this time came cotton.

## COTTON

Cotton is a presence in our wardrobes and in many other places in our lives. It is known by many names in clothing textiles, including calico, muslin, denim, chino, gingham, flannel, gabardine, seersucker, velvet and corduroy. It is also the material of canvas, gauze and cheesecloth. Some of these represent different degrees of finishing of the fabric, others different weaves, and still others distinctive dying.

Just as sheep were among the first domesticated animals, cotton was one of the first domesticated plants, in this case specifically for its fibres. Not much is known about what wild cotton looked like, though the plants likely changed into a more manageable size and became more productive of fibre, with easier harvesting, as a consequence of domestication. The most important thing to know about cotton as a crop is that it grows well only when the climate is right. Ideally it must have frost-free conditions, with temperatures of 15–25°C during the growing season, and 50–60 centimetres of rain. This means cotton is a crop for the southern US, Central and parts of South America, regions of Africa, Australia, and parts of Asia, roughly speaking latitudes of 30 south to 35 north (approximately as far south as Perth in Australia and as far north as Charlotte in North Carolina).

The story of cotton as a textile material in the context of competitive advantage is distinct from all those materials that came before or after. Growing and harvesting the fibres from a cotton boll is very labour intensive, much more so than raising and shearing sheep. Slaves from Africa were thought to be the only answer to this problem. The history of cotton textiles manufactured at scale is intimately tied to both the establishment of a large African slave population in the Americas, and to its employment. But in terms of competitive advantage, particularly for Britain, it is also the story of industrialisation, mechanisation and, as was the case with wool, empire.

Globalisation is sometimes discussed as if it was a phenomenon of the late 20th century. Of course, it is not. Humans have traded over large distances for millennia. In the ancient Native American settlement of Cahokia, near St Louis in the centre of the US, there are artefacts from both the east and west coasts. The Roman Empire had trade routes that sprawled over a very large area. But after the fall of the Roman Empire, there was an extensive period of greater localism that only ebbed away with the age of exploration and discovery during the late 15th and 16th centuries. While some individuals associated with this period were undoubtedly motivated by an overwhelming desire to know 'what was out there', the motivation of those who funded them, the European monarchs, was clearly economic: 'What wealth is out there?'

In this regard, well before Columbus, the riches to the east of Europe, of India and China, were well known. Marco Polo's accounts of these from the last quarter of the 13th century were an inspiration to Columbus and others. The dangers and difficulties faced by merchants of those times, and earlier, were immense.[9] Yet they persisted, and the goods – textiles, spices, metalwork and jewels – they brought were highly valued. For the British, the Dutch, French, Spanish and Portuguese, and selectively for other European nations, there emerged in the 16th and 17th centuries an alignment of government, business, powerful individuals in society and the military towards establishing lucrative trade routes using the import of raw materials and the export of finished goods to achieve substantial competitive dominance.

For Britain, a big part of this came to be about cotton. To exploit the riches of Southeast Asia and India through trade the East India Company was founded on 31 December 1600.[10] As the 17th and 18th centuries progressed, this company played an increasingly significant role in cotton textiles. Among other valuable products, India produced excellent woven and dyed textiles, and the East India Company tried to develop the market for these in Britain, although initially this was as a sideline to their primary business in spices. The company exerted influence during the second half of the 17th century on the Indian producers so that they would make fabrics that appealed to the British market, while also trying to influence the British consumers to like Indian fabrics. As the slave trade increased, the East India Company became increasingly involved in Africa, and used Indian textiles with colours and designs appealing to African chieftains to barter for slaves.

After 1700 this all began to change. With Indian textiles, such as calico and chintz, taking a larger share of the UK market, Parliament was petitioned to protect domestic producers, especially the wool manufacturers. The Calico Acts of 1700 and 1721 forbade the import of cotton textiles. The first act was not remarkably successful, as demand was too great, but the second act imposed stricter penalties. It did, however, exempt the import of raw cotton. In this way two things happened: India began to deindustrialise with respect to cotton, having lost one of its most important markets for finished goods. In the 1720s, Britain then began to import raw cotton and develop its own mills for making cloth. These mills did not try to replicate the family craft spinning and weaving of the Indian industry. Rather, they looked towards the successful model of the wool industry, soon going far beyond anything that wool had achieved. Thus were the seeds planted for the Industrial Revolution, and one of the greatest cases in the history of national competitive advantage.

The other major interplay between the slave trade and cotton was of course the American South. Once again, the climate was perfect for cotton growing, there was a lot of fertile land, and all that had to be put in place was a supply of labour. For about 150 years, slavery was the answer to that. In this case there was no need to deindustrialise because the South had not developed manufacturing capability, and the capability to manufacture clothing in the northern states of New England was devoted more to shoes than cotton textiles. For the South, this was the case until well into the 1800s. As a result, British cotton textile manufacture could grow with ample supplies of cotton from India, the West Indies and the US.

The growth of import of raw cotton into Britain following the Calico Laws was, after an initial burst, modest through to 1740.[11] It approximately quadrupled between 1740 and 1780, from 1.6 to 6.7 million pounds. And then by 1790 it increased to 31.5 million pounds, and by 1800 to 56 million. There are writers about the Industrial Revolution who say it was not really a revolution but something that happened gradually over a century. While there is some truth in that, the consequences of the inventions that drove the increase in cotton imports took place over a relatively short period of time in terms of economic history.

The other side of this story is the export of British-made cotton goods. In 1780 this was worth about £355,000. In 1800 it was £5.4 million. The impact on the economy of Britain can be understood by realising that exports of manufactured cotton goods grew from 5 per cent of total exports to more than 40 per cent in 20 years. But this rapid growth was only the beginning. In the 25 years from 1820 to 1845 the output of the industry grew by 40 per cent. During these decades raw cotton was about 20 per cent of British imports and finished cotton goods about 50 per cent of exports. The entire industrial base that supplied equipment to the mills, and all the shipping industry that brought raw material and took the goods away, expanded and depended upon cotton.

What about employment? In Lancashire, the county around Manchester[12] where a large proportion of the mills were located, and which increasingly was the centre of the cotton industry in Britain, 242,000 people were employed in the mills in 1801, and by 1821 that had grown to about 370,000. But thereafter, despite rapid growth of the industry there was relatively little growth in direct employment. As Hobsbawm points out,[13] from 1820 to 1845 the net output of the industry grew by 40 per cent and the wage bill by only about 5 per cent. Of course, the indirect employment consequential to this growth was very great.

### NUMBER OF LOOMS IN UK[14]

| Year  | 1803  | 1820   | 1829   | 1833    | 1857    |
|-------|-------|--------|--------|---------|---------|
| Looms | 2,400 | 14,650 | 55,500 | 100,000 | 250,000 |

How did all this happen? It is a clear realisation of the systemic routes to competitive advantage outlined in Chapter 1, and of the factors discussed in Chapter 2 that must be present in a society, in a nation, that allow the establishment of such an advantage. Research, development and invention led to successive improvements, particularly efficiencies, in the multi-stage process from cotton plant to finished goods. As each of these improvements was implemented, the industry could economically increase its scale. The lower cost of finished products, produced in quantity, led to increased demand, including domestic but especially in export markets. The huge positive balance of trade that resulted allowed for increased wealth in the society, with consequent growth in the middle class, which increased the advantage of other industries as well. This wealth also funded the growth and maintenance of the British Empire.

The invention and innovation that drove this forward were embedded in an economic system that rewarded enterprise, so there was incentive to do more. The innovations in the cotton industry occurred numerous times over more than a century. In addition to the key direct inventions,[15] such as the flying shuttle, spinning jenny and the water frame, there was the fortunate coincidence of the development for manufacture by Watt and Boulton of Newcomen's earlier invention of the steam engine. While early cotton mills were located at good sites for water wheels to supply power, this was supplanted by the steam-powered mule and loom.

All of this led to an increase in the middle class. True, there was exploitation of workers, especially women and including children. But over time fewer of them were required per unit of output, and some mill owners, particularly Richard Arkwright's Crompton Mill, developed progressive policies which laid the foundations for changes in the society for the better. In addition to reducing the number of employees required to carry out the cotton mill operations, inventions such as the self-acting mule reduced the skill level required of the worker who was operating it.

While in the early decades attempts were made through the legal system to protect the industry from foreign competition, these were largely ineffective and in the end unnecessary. Indeed, while the two biggest suppliers of raw cotton, the southern US and India, could have taken up the same industrial practices as Britain, and deprived it of its advantage, for various reasons they did not. In the case of the United States, they did not develop their own large mills until after the Civil War, by which time the British cotton industry had reaped huge rewards for the country. India, thinking that it had the lowest costs already in its family workshops, did not see the need for mechanisation. Certainly, their British rulers did nothing to encourage it. For India, the result of this was deindustrialisation, losing their once thriving textile industry based around small workshops as they exported their raw cotton.

The wealth of nations in modern times is intimately connected to the growth of great cities.[16] Just as the wool industry led to the development of Leeds as a major centre in Yorkshire, so the cotton industry in Lancashire was instrumental in the growth of Manchester. Although they are only about 50 miles apart, the climate in Lancashire is more humid, and this proved advantageous for handling cotton thread, resulting in less breakage, compared to Yorkshire. Infrastructure development, particularly canals and later railways, was again put in place to bring in raw material from ports such as Liverpool and Bristol, bring coal to feed the steam engines, and get finished goods to the ships for export. Food packaging and preservation, as discussed in Chapter 3, was also crucial to allowing for this dramatic urbanisation. Again, food packaging was a materials technology where Britain established an early advantage.

Never in the history of materials has there been such a confluence of all the requirements for a country to develop the enterprises that would achieve and sustain competitive advantage. Using its empire, and relations with former colonies in America, a spirit of invention and manufacturing, the development of large workplaces (factories) replacing small workshops to achieve scale, being better at every aspect of this than other nations, Britain's Industrial Revolution built its success in cotton textiles to achieve economic and military dominance while improving life for at least a broad swathe of its own population. Eventually, in the later part of the 19th and into the 20th centuries this would end, but it was sure great while it lasted!

## NYLON

*Gone are the days when I'd answer the bell*
*Find a salesman with stockings to sell*
*Gleam in his eye and measuring tape in his hand*
*I get the urge to go splurging on hose*
*Nylons a dozen of those*
*Now poor or rich we're enduring instead*
*Woolens which itch*
*Rayons that spread*

*I'll be happy when the nylons bloom again*
*Cotton is monotonous to men*
*Only way to keep affection fresh*
*Get some mesh for your flesh*
*I'll be happy when the nylons bloom again*
*Ain't no need to blow no sirens then*

*When the frozen hosen can appear*
*Man that means all clear*

*Working women of the USA and Britain*
*Humble dowager or lowly debutant*
*We'll be happy as puppy or a kitten*
*Stepping back into their nylons of DuPont*

George Marion, Jr. and Fats Waller, 'When the Nylons Bloom Again (a lament for the lack of Nylon stockings during World War II)'

The competitive advantage that Britain secured in wool and cotton was built in a deliberate way over centuries. Every systemic component that was described in Chapter 1 was involved, particularly for cotton. As well, the key national attributes that can lead to a sustainable competitive advantage described in Chapter 2 were present. The resulting advantage was exploited following several of the routes enumerated in Chapter 1. At least from the viewpoint of an observer two centuries later, it was a deliberate and unrelenting march to success.

Nothing could be more different from this than the competitive advantage that the United States secured through the invention of Nylon.

As discussed in Tech Talk 2, Nylon was the product of a systematic research effort at the DuPont corporation. In 1920 the company had entered the synthetic textile fibre business through acquisition of a 60 per cent share in a French rayon company Comptoir des Textiles Artificiels and mounted a vigorous research programme to see if the properties of rayon could be improved. In parallel with this, a basic research program in polymer science was begun. It was from the research group of Wallace Carothers at DuPont that the polyamide subsequently given the name Nylon[17] was developed. Throughout the DuPont research programme, the focus was on synthetic polymers that could form fibres, emphasising strength, thermal properties (so that they could be washed and ironed) and durability. Choosing starting materials for the synthesis that were readily available at low cost was also a factor.

From the start of the programme, DuPont decided to target women's stockings, as the trend towards higher hemlines meant more women were buying stockings, and that they wanted them to have an attractive appearance. Most of this market was taken by silk, and most of that was imported from Japan. The focus of the synthetic polymers programme was clearly artificial silk. The research programme was thus designed to provide competitive advantage to the United States versus Japan. While DuPont developed a textile department in the research centre, its function was to make samples to test with end users and demonstrate to DuPont's actual customers, who were the mills making the textiles and finished stockings. DuPont's product was Nylon yarn, but it had a vision of an integrated industrial system in mind. When the first stockings were introduced to the market in 1938 (but with only a small production run) they were a great success, although widespread availability happened only in May 1940. Nylon stockings became popular very quickly, and within two years of being sold were already capturing 30 per cent of the market.

It was the timing of the discovery that led to an inadvertent competitive advantage, but one with great consequences. When the US entered World War II in December 1941 most aspects of industry and research turned towards the war effort. A focus of the War Department was to look at all items that were either imported from Japan, or more broadly from the areas of Asia that Japan had conquered. Thus the US mounted a large effort on synthetic anti-malarial drugs to replace quinine that had been imported from this region, and which would be crucial for troops fighting in the Pacific Theatre. Germany similarly used its expertise in industrial organic chemistry to find ways to make fuel from coal to replace its supply of oil.

Silk too played a part in warfare, particularly as the main material for parachutes. DuPont and its customers stopped all production of Nylon for stockings until after the war and turned to the (relatively simple) production of large sheets for parachutes, which were quickly shown to have the strength and other properties that were required. This fortuitous timing of not just the invention of Nylon but the scaling up of its production allowed this materials advantage of the US in warfare to eliminate the reliance on Japan. Nylon also provided high-strength rope for many military applications. Over the course of the war Nylon was used for tyre cords, glider tow ropes, aircraft fuel tanks, flak jackets, shoelaces and mosquito netting. When DuPont returned to producing Nylon for stockings immediately after the war, they had four years of pent-up demand. They also had many other markets in which its use had been proven.

## SYNTHETIC FIBRES BEYOND NYLON

Synthetics follow a few distinctive routes beyond Nylon. The direct descendants of Nylon are the DuPont fibres Nomex and Kevlar. Their chemistry is like that of Nylon, but they have desirable properties of flame retardancy and extreme strength respectively. Another path is polyester. Carothers' team at DuPont had already made polyester fibres but set aside this research area because they found Nylon more promising in terms of fibre strength and heat stability. At the Calico Printers Association in Britain polyester chemistry was pursued and led to Terylene, then, through technical cooperation and licensing, to DuPont's Dacron. With polyester it was possible to achieve very low-cost textiles, and the beginnings of permanent press fabrics which did not crease on washing and drying. A third chemical path was acrylics, also achieving low cost and incorporating a property of elasticity. By 1965 synthetics were 63 per cent of the world textile market.

The ability to achieve vast economies of scale with synthetics, coupled with numerous advances in computer-controlled cutting and fabrication of garments, colouring and printing, led to a change in clothing in the second half of the 20th century that the world had never seen before, even with the widespread availability of various cotton fabrics. Clothing became cheap, even couturier clothing became accessible to wide swathes of society. There was a democratisation of fashion. For millennia clothing had been coincident with class, indeed in many societies wearing of certain textiles or items of clothing was restricted to members of the elite in society. Synthetics changed all of this. Because the chemistry and industrial processes involved were globally available, and research improved these in many places, the possibility of broad competitive advantage through clothing materials disappeared. There is still narrow competitive advantage in areas such as shoes or materials for sports, though these tend to last for relatively short times.

## COMPETITIVE ADVANTAGE AND CLOTHING

Synthetics declined as a share of the global apparel market after the 1960s, with a renewed interest in natural materials; these coexist, not only in the world but in most of our wardrobes. The global apparel market is of the order of $1.5 trillion. Today, there is no competitive advantage associated with clothing materials. But for the centuries of British dominance of wool and cotton, made possible by a large dose of technology combined with industrial structure, attention to infrastructure, societal economic factors, and the political will to use the Empire, especially India, for ruthless economic gain, Britain had enormous wealth that it garnered at the expense of competitor countries. The Porter factors and Landes principles described in Chapter 2 led to a positive feedback from wool and cotton that made a small country a world power. While we may never again see competitive advantage of the sort that British industrial enterprise enjoyed with wool and cotton, the materials of clothing demonstrated most clearly the principles of national competitive advantage, how it could be gained, exploited, and eventually lost.

## CLOTHING TIMELINE

| | | YEAR BP |
|---|---|---|
| -6000 | Silk production in China | 8022 |
| -4000 | Egyptian linen production | 6022 |
| -3000 | First breeding of sheep for wool | 5022 |
| -114 | Opening of the silk route | 2136 |
| 500 | Spinning wheel in India | 1522 |
| 1000 | Knitting in Egypt | 1022 |
| 1185 | Water powered fulling of wool | 837 |
| 1200 | Selective sheep breeding in Spain and England for fine wool | 822 |
| 1297 | Edward I taxation of wool | 725 |
| 1600 | British East India Company founded | 422 |
| 1614 | Henry III bans wool exports | 408 |
| 1700 | Calico Act | 322 |
| 1721 | Second Calico Act | 301 |
| 1733 | Flying shuttle (Kay) | 289 |
| 1764 | Spinning Jenny (Hargreaves) | 258 |

| | | YEAR BP |
|---|---|---|
| **1769** | Water frame (Arkwright) | 253 |
| **1774** | Leeds-Liverpool canal opened | 248 |
| **1779** | Steam-powered mule (Crompton) | 243 |
| **1784** | Steam-powered loom (Cartwright) | 238 |
| **1794** | Cotton Gin (Whitney) | 228 |
| **1873** | Blue jeans patent to Levi Strauss | 149 |
| **1892** | Rayon | 130 |
| **1938** | First Nylon stocking sold | 84 |
| **1953** | Commercial polyester (Terylene) | 69 |
| **1966** | Nomex (Sweeney, DuPont) | 56 |
| **1966** | Kevlar (Kwolek, DuPont) | 56 |

# Notes

1  Eric Hobsbawm, *Industry and Empire*, Penguin revised edition, 1999, p36

2  Kristin Romey, 'Here's What the Iceman Was Wearing When He Died 5,300 Years Ago', *National Geographic*, 18 August 2016. Accessed 16 June 2020

3  Elizabeth Wayland Barber, *Prehistoric Textiles*, Princeton University Press, 1991

4  Kassia St Clair, *The Golden Thread*, John Murray, London, 2018

5  Eileen Power, *The Wool Trade in English Medieval History*, Ford Lectures, 1941, contains a detailed discussion of the English wool industry and how it achieved European domination. The published text of her lectures can be found at https://socialsciences.mcmaster.ca/econ/ugcm/3ll3/power/WoolTrade.pdf accessed 14 June 2020

6  John Browne, *Make Think Imagine, Engineering the Future of Civilization*, Bloomsbury, London, 2019, shows the key role that engineering plays in driving societies forward. Britain's superiority in the earliest phases of industrialisation was overtaken by Germany, which built a superior academic engineering base and accorded greater prestige to engineers in society.

7  The English tried, especially in the 1500s, to both stimulate and control the domestic market for clothing through the so-called Sumptuary Laws. A 1571 Act of Parliament to stimulate domestic wool consumption and general trade decreed that on Sundays and holidays, all males over six years of age, except for the nobility and persons of degree, were to wear woollen caps on pain of a fine of three farthings per day.

8  Eulanda A. Sanders, *The Politics of Textiles Used in African American Slave Clothing*, Textile Society of America Symposium Proceedings, p740. Published in Textiles and Politics: Textile Society of America 13th Biennial Symposium Proceedings, Washington, DC, 18 September–22 September 2012 https://digitalcommons.unl.edu/tsaconf/740 accessed 15 June 2020

9  For a fascinating fictional account of merchant traders around year 1000 see A. B. Yehoshua, *Journey to the End of the Millennium*, Harcourt, 2000.

10  Also called English East India Company, formally the (1600–1708) Governor and Company of Merchants of London Trading into the East Indies or (1708–1873) United Company of Merchants of England Trading to the East Indies.

11  Most of the data cited in this section can be found in Hobsbawm, *Industry and Empire*.

12 Lancashire at the time being discussed here included Manchester and Liverpool. In the 1970s boundaries were redrawn and these cities are no longer part of Lancashire.

13 Eric Hobsbawm, *Industry and Empire*, p47

14 Richard L. Hills, *Power from Steam*, Cambridge University Press, 1993

15 See Tech Talk 3 for a description of inventions related to cotton.

16 Jane Jacobs, *Cities and the Wealth of Nations*, Vintage, 1985

17 DuPont executive Ernest Gladding, quoted in https://www.sciencehistory.org/distillations/nylon-a-revolution-in-textiles said that other names considered were Duparon (DuPont pulls a rabbit out of nitrogen) and Nuron (No run spelled backwards) before settling on Nilon, which was changed to Nylon so it was clear how to pronounce it. To make it seem more like the name of a natural material (which it was not) DuPont did not trademark the name Nylon, and often spelled it with a lower case n, nylon, encouraging its use as a synonym for stockings.

# 5. INFORMATION

On 10 May 1933, students in 34 university towns across Germany burned over 25,000 books. The works of Jewish authors like Albert Einstein and Sigmund Freud went up in flames alongside blacklisted American authors such as Ernest Hemingway and Helen Keller. The first books burned were those of Karl Marx. This book burning was in response to a speech earlier that day in Berlin by Joseph Goebbels, who declared:

> *The era of extreme Jewish intellectualism is now at an end. The breakthrough of the German revolution has again cleared the way on the German path ... The future German man will not just be a man of books, but a man of character. It is to this end that we want to educate you. As a young person, to already have the courage to face the pitiless glare, to overcome the fear of death, and to regain respect for death – this is the task of this young generation. And thus you do well in this midnight hour to commit to the flames the evil spirit of the past. This is a strong, great and symbolic deed – a deed which should document the following for the world to know – Here the intellectual foundation of the November Republic[1] is sinking to the ground, but from this wreckage the phoenix of a new spirit will triumphantly rise.*

Nonetheless, as Helen Keller said in her letter to the German students, 'You may burn my books and the books of the best minds in Europe, but the ideas those books contain have passed through millions of channels and will go on.'

The crucial material for 2,000 years of information storage and retrieval has been paper. True, it has been supplanted to a great extent by magnetic media, but the development of paper through several technological evolutions, leading to high volumes at extremely low costs, has been of enormous consequence to the development of human civilisation. Without it we would lack access to the great works of literature, we would have no instruction manuals for our machines, no daily newspapers, and when telephones became widespread, we would not have had directories in which to look up numbers of friends and businesses. Paper (and the instruments for writing and printing) facilitated literacy, and this in turn enabled almost every aspect of sustainable development described in Chapter 1. Those societies that embraced universal education leading to literacy and numeracy, a free press, and a knowledge-based society developed competitive advantage over others who saw the flow of ideas as dangerous. It was the great paper enterprises that made this possible.

From the mid-19th century until the late 20th century, film for recording both still and moving images assumed a major role in capturing and storing information in the form of pictures. The ability to reproduce photographs in newspapers made events visible to the public, especially those that were distant from the experience of much of the populace, such as front-line combat. Once again this had a profound effect on how government functioned, and where a free press was able to bring such photographs for public viewing it improved the governance aspect of sustainable development.

## PAPER

The essential material of books and newspapers, indeed of the sole source of large-scale archival information before electronic, magnetic and photographic media, is paper. Of course, there are ancient manuscripts on various forms of animal skins (vellum and parchment), indeed British laws are still archived on vellum. This is fine for single copies, and the durability of these manuscripts spans millennia. There are old Chinese writings on silk that are also well preserved. The Egyptians developed papyrus, which is closest to the idea of paper (and from which the name paper derives), slices from a marsh reed that are moistened and flattened, then can be pasted together, and form a scroll. Unlike vellum, parchment and silk, papyrus is composed of cellulose fibres (see Tech Talk 2), though they do not go through the chemical pulping process that makes paper.

It was the invention of paper – fibre from plants with water and other chemicals converted to pulp, then manufactured into thin sheets – and successive technological advances in its production which brought about one of the biggest changes in human civilisation in the past 2,000 years. It is through paper as an inexpensive material, in combination with the associated technologies for writing and printing, that literacy and numeracy became widespread. Societies that were at the forefront of the use of this material achieved the competitive advantages associated with high literacy rates. Because they valued paper, investing in its production so as to progressively lower cost, that advantage could be multiplied by developing numerous other uses for the same material.

Given that UNESCO estimates the current global literacy rate as 86 per cent, and for youth 91 per cent, it is surprising how recent an achievement this is. While historical records suffer from different measurement methods, the variability across countries is very striking, as shown in the following table[2] for selected areas in ca. 1700.

| REGION | MALE LITERACY (%) | FEMALE LITERACY (%) |
| --- | --- | --- |
| England | 40 | 25 |
| France | 29 | 14 |
| Amsterdam | 70 | 44 |
| Sweden (reading only) | 89 | 89 |
| Iceland | 50 | 50 |
| New England[3] | 70 | |

Male and Female Literacy in 1700

Even these numbers represent a considerable change from two centuries earlier, when these same regions would have had male literacy of less than 10 per cent and female literacy ca. 1 per cent. As late as 1850 only about 50 per cent of men and 30 per cent of women were able to sign their name at marriage in England.

It is generally accepted by historians that paper was invented in China, with the most documented development being that of Cai Lun in 105 CE, although there is evidence for some paper being made centuries earlier. Paper requires a source of fibre, and Cai Lun used tree bark, hemp, rags, and even discarded fishing nets.[4] Even the relatively few sheets of paper that could be made by Cai Lun's process contributed to an increase in literacy and literature in ancient China. There are records showing that scholars (a distinct group in China) were issued with a thousand sheets of paper each month. In turn this required a step up in the manufacture of brushes and ink for calligraphy. It was not until about the 8th century CE that block printing was invented, either in China or Korea, and this again led to an increase in the demand for paper.

## The Islamic Golden Age

As with silk, the Chinese were very secretive about paper. Undoubtedly, knowledge of how to make it spread westward via the Silk Road. Legend has it that during the Battle of Talas (Kyrgyzstan in Central Asia) in 751, in which the Muslim Caliphate won a decisive victory over the Chinese, a number of Chinese prisoners were told that they would be executed unless they could teach their captors something interesting, and one of them volunteered 'I know how to make paper'. One can only imagine the dialogue that ensued. Whether this revelation of the secret process or a more evolutionary spread of the technology was what actually happened, the methodology to make a watery fibrous pulp, then press it on screens and produce sheets of paper, rapidly became known and used in the Arab world. Under the Abbasid Caliphate (750–1258 CE) papermaking was established in Samarkand (Uzbekistan), then in Baghdad, Damascus and Egypt.

It was this command of the technology for making paper at scale that enabled the so-called Islamic Golden Age. This period, from ca. 790–1300 CE, saw the translation of numerous classical works of civilisations that had been conquered by the Muslims from Greek, Persian, Chinese, Sanskrit and other languages into Arabic, Syriac, and later into Turkish, Hebrew, Persian and Latin. This was not just a preserving of knowledge of these cultures; because of the availability of paper it was a propagation of it across a wide geographic region, from Central Asia to

Spain, and then through centres of scholarship, leading to a vast extension of the knowledge encompassing mathematics, philosophy, law, the physical sciences, astronomy, medicine, engineering, and even social sciences such as economics. The educational institutions established during this period, the madrassas, certainly had religious teaching at their centre, but taught boys and young men many secular subjects as well. At that time Islam saw education (of males) as something that was commanded by their religion.

To facilitate this, paper production was established in Baghdad, with linen rags the primary source of fibre. Enormous quantities of rags were needed to provide fibre for paper production, a practice that would dominate paper until the second half of the 19th century. The Muslim leaders saw early on the great superiority of paper over papyrus or parchment. This was not just in its physical strength and durability, but in its way of binding ink to make erasure difficult. The value of this is clear where paper is being used to keep records. It was during these centuries of the Islamic Golden Age that the practice of paper-based record keeping of all sorts – production, debts, taxes, land titles – became established, enabled by the ability of each party or government office to hold copies. There was an extensive merchant class (very much the forerunner of the middle class created by the Industrial Revolution) in this society, and its success was facilitated by paper records. In this way, the paper enterprise became a key component of the growth of numerous other enterprises.

Printing at any scale did not yet exist, of course. But the great centres like Baghdad set up production lines for hand copying of books to produce large numbers of copies. While the core motivation was religious, this became a small part of the dissemination of knowledge through books. For the first time in human civilisation, about 1,000 years ago, someone could earn a living by writing books. While ancient libraries, for example in Alexandria, were places to hold collections of manuscripts, the public lending library was introduced for the first time in Baghdad and later in several other urban centres. Bookbinding became an important professional skill, with distinctive forms of paper products required to provide durable binders.

The material we know as paper thus secured competitive advantage for the Abbasid Caliphate for more than 500 years. It began the democratisation of literacy and the spread of knowledge. Medical texts, many drawing on Greek science from antiquity, improved health care over a vast geography. Teaching mathematical principles from books led to a change in numeracy never seen in the history of human civilisation, while both knowledge of fundamentals and the advances in mathematics occurring in this society led to superiority in engineering as well. Through its multiple uses by the merchants, paper provided a sound foundation for business transactions. Even so, this was but the first big step for paper as a source of national competitive advantage.

## Western Europe and America — The Rag Era

> RAGS *make paper*
> PAPER *makes money*
> MONEY *makes banks*
> BANKS *make loans*
> LOANS *make beggars*
> BEGGARS *make RAGS*
> Anon. 18th century[5]

Paper spread from Arab Spain into the rest of Europe. In the 1200s there was papermaking in Italy, and throughout the 1300s mills were built to make paper in Germany, France, and the Netherlands, Switzerland in the 1400s, and eventually to England, though the first commercially successful mills in England were not until the 1500s. All these mills relied on rags for fibre, and since wool fibre, being protein rather than cellulosic,[6] is not suitable for paper, this was mostly linen until the growth of cotton clothing. Many historians have remarked on the growth of paper production in Europe being a stimulus to the invention of movable type and Gutenberg's printing of the Bible in 1452,[7] and that growth in turn having been made possible by the earlier advances in spinning and weaving that increased the supply of cloth. Well into the 1800s these two industries, clothing and paper, would be close coupled, until wood pulp displaced linen and cotton as the source of fibre. Rags to paper was the largest recycling (or better put, repurposing) materials industry the world would see until well into the 20th century.

Printing changed everything. In the 50 years following the Gutenberg Bible ca. 40,000 separate titles with 15 million copies were printed in Europe. All of this required increased quantities of paper. The first paper mill in America was established in Philadelphia in 1690 by William Rittenhouse. In the 18th century mills proliferated in New England. Once again, the paper was made from rags, predominately cotton and linen imported from Europe. Competing demand for rags between the different paper manufacturing centres was great. In the 1850s rags represented half the cost of making paper in the United States.

There are several other discoveries and inventions that occurred during this era of European/American paper. Two of them concern writing rather than printing. In about 1660 the first fountain pen was made, a writing nib with its own ink reservoir. It took many decades to perfect this, and even longer for it to displace the quill and inkwell. A century earlier, in Cumbria, England, a large deposit of graphite was discovered. While people had been using lead as a soft metal for writing, graphite quickly proved much more useful. Graphite sticks encased in wood were available from the early 1600s, but in the 1660s, at about the same time as early fountain pens, mass-produced pencils were manufactured.

The third major development related to the chemistry of rags to paper, namely, the discovery of chlorine bleaching. German-Swedish chemist Carl Scheele, who discovered many of the chemical elements, was the first to isolate chlorine gas in 1774, and there followed a series of developments of chlorine for bleaching. Strong bleaching of dyes from rags increased the supply that was suitable for use in paper.

How did competitive advantage emerge from this revolution in printing of books in the 15th to 18th centuries? Europe in these centuries was in religious turmoil, the ramifications of which are still being felt today. Martin Luther nailed his 95 theses to a church door in 1517, and Henry VIII broke with Rome in 1534. In the same period, the Catholic Church in Iberia was becoming stricter and more repressive of dissent. This was also the age of exploration. While Spain and Portugal were leaders in this in the 16th century, the English, French and Dutch played major roles in the 17th. One of the earliest uses of printing was maps, with the first printed map probably dating from 1472, and subsequent maps documenting the changing understanding of the world. Increasingly detailed printed maps, available at lower and lower cost, eventually given away for free at filling stations in the 1950s, were an important advantage to those who had access to them, resulting from the availability of cheap paper.[8]

One of the ways these differences were to manifest themselves in the 17th century was in attitudes towards literacy. The Church in Southern European countries resisted growth in literacy that would enable large numbers of people to read the Bible themselves. Ironically, much of this took place in the very region where the Islamic Golden Age occurred just a few centuries earlier. In reaction and opposition to this, northern countries and the settlements in America, especially in New England, began to push for universal literacy. Most notable were laws in Sweden (and Finland, which was a Swedish province at this time), first in 1571 and more comprehensively in 1686, that effectively made it compulsory for everyone to be able to read the Bible and answer basic questions about it. Because of these laws all Swedish cities, towns and villages were required to organise schools for reading and counting. Sweden was distinctive as well in achieving comparable or even higher literacy rates for women as for men. In 1642 the Massachusetts Bay Colony passed a law requiring that all children be taught to read and write. The Puritan settlers believed that the success of the colony was dependent on everyone being able to read the Bible and the laws of the land. This law was followed by another in 1647 requiring the establishment of public schools. To be sure, these were not usually free of charge, were often not open to girls, and, given the agricultural society, it did not lead to universal literacy. But the principles were established and spread throughout the north-eastern colonies. The high literacy rates shown in Table 5.1 show the differentiation that occurred between nations or regions. This differentiation remained very pronounced as late as 1900, with Britain having a literacy rate of 97 per cent, contrasted with Italy at 52 per cent, Spain at 44 per cent, and Portugal 22 per cent.[9]

It has been asserted that literacy does not afford competitive advantage. In a society where there is only very limited literacy, there may be evidence that the literate and numerate individuals do not necessarily advance economically compared to their illiterate counterparts.[10] One of the assertions is that, for example in Sweden, there is evidence that reading comprehension was far below the ability to just read a page of text. While this may be true for the Bible (one can imagine the priest enquiring of a parishioner, 'you just read the sentence, "Vanity of vanities, all is vanity", please explain what that means') the same person, reading the sentence in an instruction manual for a piece of machinery 'turn switch to position 3, then lift lever to grind corn', will know just what to do, whereas an illiterate counterpart will require each step to be explained and memorised.

Investment by a nation to achieve high levels of literacy secured competitive advantage that lasted for generations against other nations that did not see its importance. Crucial to this was the availability of books at all levels, i.e., not just scholarly works but primers that could be used to instruct children, and to instil in them a love of reading. Paper was also important for books to teach arithmetic, again securing the advantages inherent in a broadly numerate population. Even in societies where women played a small role in the workplace outside the home, their literacy was important in starting the next generation on the right path. This was particularly the case in agricultural regions where public schooling was not available until late compared to urban areas.

It should not be assumed that where literacy rates were comparatively low (for example England compared to New England) this was completely related to historical class position. There emerged in England and in other mercantile countries of Europe such as the Netherlands a class of clerks, accountants, lawyers and shopkeepers who were literate and numerate. They were an effective middle class and were an important part of the success of these nations over several centuries. The ability to do their work depended on literacy and numeracy acquired as children, the skills required for the work were usually acquired through apprenticeships, and the work itself was paperwork!

## Trees to Paper

The demand for paper was, of course, not just for books, but also for newspapers, even before many of the other uses of paper were developed, although niche uses such as hats and collars were already taking some paper in the 1700s. It was impossible to satisfy the newspaper requirements using rags as the only source of fibre. The demand in America accelerated very rapidly – in 1843 2 million pounds of rags were imported into the US, in 1850 it was 21 million pounds, and by 1857 more than 44 million pounds.[11]

This fibre shortage was exacerbated by major advances in paper production. For centuries, paper was produced as single sheets, increasingly large sheets which could be cut to size, but still single sheets. These single sheets needed to be fed into the printer. Moreover, the production of paper as single sheets was very labour intensive. The answer was in something resembling a conveyer belt, where wet pulp was poured onto a screen at one end and in a series of steps became a continuous sheet of paper at the 'dry end'. This could then be rolled up and fed into a 'fast' printer. The first such machines are generally known as Fourdrinier machines, after two brothers, Henry and Sealy Fourdrinier, who were the first to commercialise them in England, though the original invention was French.[12] In the early 1800s machines were already well advanced in transforming the textile industry; the same technological skills base in England and Scotland evolved the continuous papermaking machines. It was a period of invention, patenting, inventing something better, as well as theft of intellectual property and the key people needed to make the machines work. The papermaking technology spread to the eastern United States in this way. As had been the case with textiles, Britain dominated the production of the machines, so that even those used in France and Germany in the first quarter of the 19th century were all British. As with textiles, it was the ability to build large enterprises that differentiated Britain. Workers revolted against the machines, but once again they lost.

The system of raw material to printed word was unblocked at the back end of the process, with fast production of continuous rolls of paper and higher-speed printing, but was severely constrained at the feedstock front end. One of the most consistent lessons in the history of innovation is that a bottleneck such as this leads to invention. Bottlenecks do not last. A new source of fibre, and a lot of it, was required.

From the mid-19th century all sorts of waste materials were used. Any textile mill, whether linen or cotton, invariably had fibrous waste. Cotton gins, used to extract the fibre from the bolls, also had piles of waste. All of this began to be incorporated into the pulp mills to extend the rag supply. Many attempts were made to use different sorts of straw or grasses, but none of these produced quality paper, probably because of limits to the chemical knowledge for processing them into pulp that could have separated out unwanted components.

Cotton is cellulose, a simple sugar polymer, but it was not at all obvious to chemists in the 19th century that trees were also more than half cellulose. Indeed, there is some variability with species, but (on a dry weight basis) most trees contain 65–75 per cent cellulose. The remaining 25–35 per cent of the tree, the largest component of which is a polymer called lignin, gives the tree its hardness, and it was this hardness which made it unobvious that a tree could contain the same polymer as a soft material like cotton. It is not surprising that the already advanced chemical establishments in Germany, and only slightly later, in the United States, were able to successfully attack the problem of producing a suitable pulp for paper from wood in the second half of the 19th century. Germany had lagged behind Britain in invention of machinery, but it built a formidable research establishment that was able to overtake the British culture of invention to produce new chemical processes. The United States in the late 19th and early 20th century drew on its British roots to build new businesses based on mechanical and electrical inventions but admired the German scientific establishment and its role in advancing chemistry and physics, both for basic understanding and for translating that into commercial success. Many of the leading US chemists spent time studying and doing research in Germany.[13]

Tech Talk 4 describes the innovative processes used to produce high-quality wood pulp for paper manufacture, once again a series of chemical developments that made increasingly better paper – whiter for printing, stronger for packaging, longer lasting without yellowing or breaking down where books were being printed, cheaper for newsprint. These processes were implemented widely first in the north-eastern US and Canada, where forests could provide huge quantities of wood, and in Wisconsin and Michigan, also good forest country and near to markets in Chicago and other growing Midwestern cities. Later these processes spread to the pine forests of Georgia in the south-eastern US and to Scandinavia. With this, the use of rags declined; paper manufacture was dominated by wood as the raw material. Rivers in these woodlands also provided the quantities of water needed for the paper mills, sometimes being badly polluted by the waste from the process. Today it is only boutique, expensive papers that are promoted as being linen or rag paper, as well as special paper for certain currencies, particularly the US dollar.

## NEWSPAPERS – A FREE PRESS AND COMPETITIVE ADVANTAGE

The emergence of competitive advantage for the United States from this development of wood pulp-based paper comes directly from the systems thinking and sources of advantage described in Chapter 1. Initially, as one would expect, numerous small- to medium-sized paper mills developed in New England, especially in Maine, which has trees and strong-flowing rivers to provide the substantial amounts of water required for the process, as well as in Canada. From these small businesses three major companies emerged: the largest was International Paper, formed from the merger of 18 separate paper companies in 1898, many of the component companies themselves having resulted from mergers of still smaller companies. The other two were Kimberly Clark, formed in 1872 as a rag paper company in Wisconsin, but quickly adapting to wood pulp, and Scott Paper Company, formed in 1879 in Philadelphia initially using mills in Nova Scotia for its paper products. All three were to have an enormous influence on American competitiveness in the 20th century.

From the earliest times of the American colonies newspapers played a key role in society. John Peter Zenger, a German immigrant, published a paper called the *New York Weekly Journal* from the 1720s onward, critical of the British governor. Eventually the governor sued him for libel and Zenger won, becoming a figure symbolic of freedom of the press. When some decades later the US Constitution was agreed, the first item in the Bill of Rights included 'Congress shall make no law … abridging the freedom of speech, or of the press'. One of the key principles of sustainable development was thus enshrined in law.

To be effective, newspapers must be cheap and have broad reach in society. Given that they will be only partially objective, there must be sufficient competition for differing views to be represented. To secure both the low cost and broad reach, in the earliest days of the United States as a nation (1792), laws were passed that required the post office to deliver newspapers at very low cost: one penny if within 100 miles, and 1 ½ cents beyond that distance (later revised to one penny anywhere within the same state). This was well below the cost of letters, though it was still a significant part of the cost to the reader, which was usually about four cents for the paper.[14] These rates remained in effect for half a century.

The question of competitive views took care of itself, as political parties emerged early. Recall that even in early US history, there was a high degree of literacy (among white men) and the political activism out of which the country was born led to a hunger for newspapers. By 1850 there were 400 daily and 3,000 weekly newspapers in the US.

The hunger for information was there, the commitment to a free press providing information as well as conflicting views, able to be critical of government and business, was also there. But the presses were not fast enough, and the supply of paper was neither sufficiently plentiful nor sufficiently cheap. All that changed with paper from wood pulp. International Paper initially concentrated its business on newsprint, integrating from forested land through paper production. By 1910 there were 2,000 daily and 14,000 weekly newspapers published in the US. International Paper supplied 60 per cent of the newsprint. Concurrent with the increase in newspapers was growth in advertising, meaning a sophistication in both typography and graphics to appeal to customers. Most of the revenue for the newspapers came from advertising, and people received most of their information about new products as well as current events from newspapers in the years before radio became widespread.

## FILM AND PHOTO REPRODUCTION

The use of photosensitive plates whose images were reproduced in print media dates back to the period around the US Civil War, though because exposures were so long, dead bodies were the best subjects, being able to remain still enough to give a clear image. Matthew Brady became synonymous with Civil War photography, and many of his (and his paid associates') photographs are in the National Archives in the US. Glass plates were only superseded in 1884 by George Eastman's invention of the first photosensitive flexible film, and then in 1889, building on chemistry developments from the 1860s, a chemist working for Eastman developed a film that could be used for negatives based on nitrated cellulose, effectively the same material base that remained throughout the next hundred years of film cameras (except for Edwin Land's Polaroid process). A novel material, built into a large commercial enterprise – the Eastman Kodak Corporation – transformed the conveying of information, this time in the form of pictures. The two key research areas after the first materials invention were understanding the chemistry of photosensitive materials so that sensitivity could be increased and exposure times decreased, and finding a way to record and reproduce full colour images.

The timing of the commercial availability of photographic film, and the cameras using it, was thus perfectly coincident with the development of paper from wood pulp. One other key invention to complete this systemic change was Frederic Ives' development, also in the 1880s, of what is known still as screen printing of halftones, halftones being the various shades of grey in a black and white image. While many others had been working on methods for doing this, Ives produced a metal plate which was divided into photosensitive dots, so that when a photo was re-photographed onto this plate, the image could then be printed. This replaced many laborious and often not very easily conducted processes for reproducing photographs in newspapers and magazines, immediately cutting the cost by 95 per cent while improving quality.

And then there came movies!

## MOVING PICTURES

Still photographic images had a profound impact, from the mid-1800s onward, on the kind of information people could receive, distribute and archive. Newspapers and magazines were powerful means for communicating information in the form of pictures. Through advertising, they drove demand for products of all sorts. While the ability to make moving pictures was, for many years, confined to a relatively small group in possession of the necessary equipment, the distribution of these, initially in music halls and soon through cinemas, meant that their impact on society was probably as great as for still images from early in the 20th century, in those countries where the business of cinemas existed and people could afford to go to them. The newsreel, a short film on current events, was introduced by Frenchman Frederick Pathé in English music halls in 1897, and from the early 1900s until sometime in the 1970s newsreels were an important means for keeping the public informed. Such was the hunger for news in this format that there were theatres established only showing newsreels.

Early in the life of the newsreel, governments realised that what was shown, or not shown, was an effective means of propaganda. This was used extensively during World War I. Even before 'talking pictures', it was possible to have a live or recorded narrator give a voiceover to the movie. The news became what the editors chose to show, and the message behind the images what the narrator said it was. By the 1930s, Hitler recognised the power of movies, both news and stories, for building alignment behind his world view.

A few countries dominated the making of movies for entertainment over many decades, with the huge enterprises built by individuals such as Samuel Goldwyn, Jack Warner, and J. Arthur Rank. Their movies were shown, and became known, around the world. They established a view of what was culturally correct and desirable and the norms of behaviour. As such, movies were a powerful form of cultural imperialism that created a demand for products from countries such as the United States, the UK, France and very few others.

Moving pictures also played a role in science. While the microscope had been invented in the 1600s, the number of people who had seen an image of living microscopic creatures was tiny at the start of the 20th century. The French filmmaker Jean Painlevé (1902–1989)[15] had the idea of making movies through a microscope. He became famous for taking samples along the shore of the sea, then making documentaries showing the movement of living creatures found there. Through more than 200 documentary films, during which he developed time lapse, slow motion, and other techniques, he 'democratised' science. His work was so astounding that it also appealed to the leading Surrealist artists of the time. In later years, Painlevé used his films and proficiency in special effects to explain such things as quantum mechanics and relativity.

## COMPETITIVE ADVANTAGE THROUGH FREE INFORMATION

One of the routes to competitive advantage described in Chapter 1 is living according to the principles of sustainable development, of which the relevant one connected to paper and other materials media is the transformation of democratic societies by the nearly free (both in terms of ideas and cost!) availability of information to citizens and the accountability to which governments and businesses were held by newspapers. At the same time, as discussed in Chapter 2, competitive advantage comes from the inexorable growth of an informed middle class, and once again the newspapers played a key role in this in the United States by conveying, at extremely low cost, news, opinion and advertising.

Now, of course the US did not have a monopoly on newspapers, although it was unique in the absoluteness of its constitutional guarantee of a free press, and protection of newspapers against libel when they printed the truth. By contrast, the freedom of expression guaranteed by the French Revolution in the document 'Declaration of the Rights of Man and of the Citizen, 1789' states:

*Liberty consists in the freedom to do everything which injures no one else; hence the exercise of the natural rights of each man has no limits except those which assure to the other members of the society the enjoyment of the same rights. These limits can only be determined by law.*

The law which set these limits specified that freedom of the press did *not* extend to

1. *Provocation to crimes or misdemeanors*
2. *Provocation to a certain number of serious crimes, in particular to attacks on the authority of the State*
3. *Seditious shouting or singing*
4. *Provocation to members of the military to turn away from their duties*
5. *Offence to the President of the Republic*
6. *Publication of false news that have disturbed the public peace*
7. *Offence to public decency*
8. *Defamation and insult*
9. *Offence and contempt towards heads of State or foreign diplomatic agents*[16]

Some of these limitations on freedom of expression in France still exist, and others were only repealed as recently as 2013.

Beyond the constitutional guarantee of freedom of the press, the competitive advantage of the United States came through a fully integrated system of natural resources in terms of forests (initially cleared but quite soon managed more sustainably), continued technological development of paper production from wood through research and engineering; the building of world-scale commercial enterprises such as Scott Paper, International Paper and Kimberly Clark; attracting some of the best minds to the leadership of newspapers (from Peter Zenger in colonial times, to Noah Webster, Horace Greeley and many others); as well as business/political figures such as William Randolph Hearst and Joseph Pulitzer,[17] as publishers. Any part of this could, and was, replicated elsewhere, in England, or in Sweden for example, but not the entire system. That was the first big paper-related and film-related source of competitive advantage for the US in the first half of the 20th century.

## COMPETITIVE ADVANTAGE THROUGH MULTIPLE USES

As discussed in Chapter 1, there is a virtuous cycle for materials where increased scale lowers costs, and lower costs mean more uses for the same material. Companies that are major manufacturers of materials, at least those that survive and prosper, have vigorous research and development programmes. These drive the cost down and reliability up for transformation of raw material to manufactured product, but they must also be sensitive to customer needs that the product could fulfil, producing modifications, sometimes slight, sometimes profound, that open new markets. With paper, there are giant mass markets like newsprint and low-cost books where supplier margins are small. But a great enterprise sees new markets that can be accessed where volumes may be (initially) smaller but value is higher.

This was the case for the dominant US paper companies, International Paper, Scott and Kimberly Clark. Their research departments knew everything they could know, given the state of chemical knowledge and instrumentation, about cellulosic materials. One of the things they were interested in from the outset was the absorption of liquids by paper. This was critical for understanding the interaction, both initially and over time, of ink with paper. But it also led to several new products.

At the beginning of World War I, Kimberly Clark research developed a process for taking cellulose from which lignin had been removed and making it into a highly absorbent material for liquids. They encased this in gauze (a cotton product, so also cellulose), and bandages were created that replaced much less effective cotton cloths. They proved this in Chicago hospitals and were able to supply large quantities to the US Army and the Red Cross when the US entered the war in 1917.

When the war ended in November 1918, the company had far more capacity to produce this packaged liquid absorber (mainly for absorbing blood) than was needed. A marketing specialist at Kimberly Clark looked for new uses and learned that nurses during the war had used these new bandages as menstrual pads. By 1920 the euphemistic term 'sanitary napkin' had been coined, and one of the most famous brand names, Kotex, had been registered. Over the coming two decades this product laid the foundations for a transformational effect on US society, because women did not have to use cloth pads that they laundered, enabling them to manage their periods more discreetly and effectively. This opened opportunities in the workforce for women, proceeding slowly to be sure during the depression but having profound consequences during World War II.

The absorbent material used for bandages was also used during World War I as a filter in gas masks. Again, the research department began to look for other uses. A paper tissue product was created for removing makeup, replacing cotton balls. But customers with hay fever found that this tissue was an effective replacement for the cloth handkerchief, and Kleenex was born as a product. It was only 10 years after the World War I armistice that the pop-up tissue box was invented.

The biggest impact in paper products came from toilet roll. While people had been using paper of various sorts to perform the wiping function, perforated absorbent paper was patented first in 1871, and then again in 1891 in the form of the roll. This invention was the basis for starting the Scott Paper Company, its use growing much more rapidly in the US than elsewhere. Later, as with tissues such as Kleenex, the idea of combining two different papers, one for softness and the other for strength, improved the efficacy of toilet paper for stopping the transmission of bacteria and viruses in faecal matter to the mouth, especially through food. Surprisingly, there is no agreement about the importance of this to the reduction of diseases such as cholera and dysentery, probably because of other hygienic practices that grew in the same period, and the widespread contemporaneous chlorination of water in the countries where this reduction occurred. But in World War II, and in the Korean War, American soldiers all carried with them their own rolls of toilet paper, and the incidence of faecal-related disease for them was markedly lower than for allies or opponents where this was not the case.

These commercial developments – Kotex, Kleenex, Scott's Toilet Tissue – represented a source of competitive advantage to the US that stemmed from late 19th century inventions, and were built to commercial scale in the period between the world wars and had a real impact on the success of the US Army and its supporting personnel during World War II. They were a consequence of the information advantage the US paper industry provided through newspapers and are a clear illustration of the systemic way a successful industrial enterprise works to drive advantages that were described in Chapter 1.

And a footnote, perhaps less persuasive in terms of national competitive advantage, on film. Roentgen discovered X-rays in 1895, and he immediately realised the potential of this for medical purposes. For instant viewing this was done in the way the original discovery was made, using a fluorescent screen that displayed an image when a subject, such as a patient with a broken bone or a bullet lodged internally, was irradiated with an X-ray source. But for many purposes, including archival records, film recording of X-rays was used. As the medical speciality of radiology developed, film became more important, as the images were taken by one expert, the radiologist, and conveyed to another, the surgeon, for treatment.

There were some rudimentary attempts to use X-ray technology in support of battlefield medicine as early as 1905, and the Germans had some mobile units that were horse drawn with generators on board. Still, at the start of World War I, X-ray machines were only available in hospitals, and mostly in urban centres. Marie Curie realised that these were needed to save the lives of wounded soldiers and developed a mobile X-ray unit in a vehicle. The engine of the car was used to provide electrical power for the X-ray generator, and there was a darkroom on board for developing films. The first machine was used as early as 1914, and through her fundraising efforts Marie Curie was able to deploy more than 20 such mobile units early in the war. The mobile nature made it possible to have the machines available as the front moved. Later, she saw to the installation of units in more permanent field hospitals. With her daughter Irene she recruited and trained the technicians to operate all the units. It is estimated that the total number of wounded receiving X-ray examination as part of their treatment during the war exceed one million.[18]

In addition to the medical applications, from the earliest days of X-ray technology, industrial applications were recognised as important. X-rays have been used continuously over the last 120 years for all sorts of non-destructive testing (NDT), for example looking for cracks in key components as they are coming out of the manufacturing process. Again, film provides the archival record that such testing took place and is thus important in disputes that may arise when a failure has occurred.

While film has all but disappeared as the medium for recording images as personal or news photographs, and digital imaging is certainly important for industrial NDT, film remains an active business for companies such as Kodak for both medical and industrial NDT. As recently as 2019 Kodak launched a new line of NDT film products. The successful aerospace, automotive and many other manufacturing industries of the US, Germany and Japan rely on NDT, and film is a part of that reliance.

## INFORMATION BEYOND PAPER AND FILM

The conveyance of information has been transformed many times over in the 20th and first two decades of the 21st centuries. Development of the science of electronic properties of solid materials growing from quantum mechanics in the early 20th century eventually led to electronic devices based on transistors (Tech Talk 5). Because these technologies relied on fundamental understanding of solid-state physics, they emerged in the second half of the 20th century from laboratories in countries such as the US, UK, Germany, France and Japan that emphasised excellence in science and engineering.

The most basic material at the heart of these devices is ultra-high-purity silicon. By high purity here is meant silicon which is 99.9999999 per cent pure as a minimum! Tech Talk 6 describes how this is accomplished. At the heart of it is a process invented by Siemens in Germany and initially licensed to companies in the US, Germany and Japan. As a result, only those countries were able to develop the technology to produce the computer chips that are core to processing and storing information. These countries had sufficient capability to build the big enterprises around high-purity silicon that led to the explosive growth of computer manufacturing in the second half of the 20th century. But the technology could not be protected forever, and as demand grew, the Chinese developed expertise and the ability to make this material at huge scale, so that today China dominates silicon production. Moreover, the same high-purity silicon is the core component of the photovoltaic cells used for solar power. As late as the year 2000, the solar business was so small that it relied on waste material or surplus from the electronics industry for its material. Once again China saw that building huge enterprises to produce high-purity silicon for solar cells was an opportunity to dominate this industry, given that the cost reductions inherent in increased scale would fuel demand. Once again a classic case of the systems diagrams shown in Chapter 1.

Radio, television, satellite transmission of images, opening of multiple television channels, personal computers, newspapers with photographs of events that occurred hours earlier on the other side of the world, everything about the internet and its interaction with every other medium of information transmission and retrieval, have been a more radical transformation in information than paper and film. For most of the world, none of these developments has led to the sort of national competitive advantage described earlier in this chapter, though they have certainly contributed to the wealth of China and Taiwan. They all involve an array of specialised materials, electronic, magnetic, phosphors, and more, but with rapidly decreasing cost so that the materials could not form the basis for competitive advantage.

Still, some nations manage to be disadvantaged around these media. As in earlier repressive regimes in the era of paper, from Catholic Spain to Hitler's Germany and Stalin's Russia, there are still nations that block the transmission of information, that stop access to the internet, and that censor what their citizens can hear or read. These countries have always disadvantaged themselves economically.

All the modern media for information are electronic and magnetic in one way or another, and for the approximately 1 billion people without access to electricity, as well as another 2 billion whose electricity supply is intermittent and unreliable, it is impossible to compete in the modern world. How the inequality of electricity came about, in particular the materials aspect of this, is discussed in Chapter 7.

Electronic materials for information storage pose a new problem and a potential for disadvantage that was not the case with paper and film. While we can still read scrolls that are 2,000 years old (for example the

Dead Sea Scrolls), it can be difficult to find a computer to read a floppy disk from 1990. Just having a disk drive that works is not sufficient either, as one needs the software to read and display the stored information. This is a recognised problem, and a Carnegie Mellon University project seeks to preserve hardware and software for older storage formats,[19] but the problem is already a massive one. Businesses, governments, museums, all need to consider the durability and accessibility of material archived using electronic materials.

There is one other key material of modern information transmission, and that is optical fibre. Developments in computers, lasers and telecommunications in the 1960s made it clear that transmitting information via copper wire would not have the capacity and speed required. Optical fibres had been known for a long time, and the idea of modulating a light signal down a fibre to transmit information was discussed, but the losses through the fibres were too great. This problem was solved following a long research effort at Corning Glass in the US by Doctors Robert Maurer, Donald Keck and Peter Schultz. They reduced the losses through fibre by the amazing factor of $10^{98}$. Once breakthrough had been achieved, improvements continued in the following decade. As with other revolutionary developments, there was a confluence of inventions and vision that changed how information was transmitted. The semiconductor laser, a solid-state device that could convert electricity to light at room temperature, was realised in about 1970 and became the ideal light source for fibre optics. Microprocessors, effectively the entire central processing unit of a large computer on a single chip, became commercially available in 1970 and allowed devices to both encode information for transmission and decode the signal received. The entire vision of an interconnected world replacing the telephone systems of old was developing at the same time.

As devices became cheaper, speed became a source of competitive advantage. Today, countries with high internet speeds, made possible by optical fibre coupled with efficient network architecture, are advantaged in running their own businesses and in attracting foreign investment. A country does not need to have the biggest or most powerful army, the most territory, the biggest population, to be successful in this information world. But it does need to have high-speed internet. A survey of the top 15 in 2017[20] is shown in the following table, which can be compared with the world average of 7.2.

| COUNTRY | AVERAGE CONNECTION SPEED MB/SEC |
|---|---|
| South Korea | 28.6 |
| Norway | 23.5 |
| Sweden | 22.5 |
| Hong Kong | 21.9 |
| Switzerland | 21.7 |
| Finland | 20.5 |
| Singapore | 20.3 |
| Japan | 20.2 |
| Denmark | 20.1 |
| US | 18.7 |
| Netherlands | 17.4 |
| Romania | 17.0 |
| Czechia | 16.9 |
| United Kingdom | 16.9 |
| Taiwan | 16.9 |

## TECHNOLOGY CHANGES, PRINCIPLES ENDURE

The transformation of information storage and retrieval in the late 20th and early 21st century seems so completely revolutionary that it is tempting to think that the principles of competitive advantage discussed in Chapters 1 and 2 must have been overturned as well. This is not the case.

Sustainable development remains a central organising principle, and one of the important components of it is a free, open and just society incorporating good governance. Without access to information – be it in paper or electronic form – citizens cannot exercise their role in this governance.

A robust infrastructure, evolving and well-maintained, will always be characteristic of a society that is competitive. This is equally true of optical fibre, high-speed telephony, and the electrical network to support their operation, as it is for bridges, roads and tunnels.

The developments that have revolutionised information have come from a few countries, fewer than 10 per cent of the members of the United Nations. These countries have well supported universities, substantial commitments to research and engineering, they protect inventors and reward innovation. Some of them are better at this than others, but all of them are better than the other 90 per cent. Once again, the principles outlined in Chapter 2 endure.

## INFORMATION TIMELINE

|  |  | YEAR BP |
|---|---|---|
| -3000 | Papyrus plant used for writing in Egypt | 5022 |
| -200 | Parchment used in Turkey | 2222 |
| 105 | Invention of paper in China by Cai Lun | 1917 |
| 800 | Papermaking in Samarkand | 1222 |
| 911 | Moors introduce paper to Spain | 1111 |
| 794 | Papermill in Baghdad | 1228 |
| 932 | Wood block printing China | 1090 |
| 1151 | First European papermill in Spain | 871 |
| 1298 | Marco Polo reports paper money in China | 724 |
| 1309 | Paper first used in England | 713 |
| 1452 | Gutenberg Bible | 570 |
| 1495 | First papermill in England | 527 |

| | | YEAR BP |
|---|---|---|
| 1564 | Graphite discovery in England | 458 |
| 1600 | Spain outlaws papermaking in its colonies | 422 |
| 1611 | King James Bible | 411 |
| 1660 | Fountain pen | 362 |
| 1662 | Mass produced wood/graphite pencil | 360 |
| 1757 | Woven paper by Whatman | 265 |
| 1803 | Fourdrinier continuous papermaking machine | 219 |
| 1820 | Photoengraving | 202 |
| 1826 | First photograph (Niepce) | 196 |
| 1830 | First paperback books published in England and Ireland | 192 |
| 1837 | Multicolour printing | 185 |
| 1826 | Fanshaw production printing New York | 196 |
| 1839 | Negative positive photography (Fox-Talbot) | 183 |
| 1841 | Voigtlander camera | 181 |
| 1848 | Photojournalism of French revolt | 174 |
| 1854 | Photojournalism of Crimean War | 168 |
| 1863 | Sulphite process for paper from wood pulp | 159 |
| 1874 | Remington No. 1 typewriter | 148 |
| 1876 | Bell's invention of the telephone | 146 |
| 1879 | Sulphate process for papermaking | 143 |
| 1877 | Edison invention of the phonograph | 145 |

| | | YEAR BP |
|---|---|---|
| **1887** | Roll celluloid film | 135 |
| **1888** | Kodak camera | 134 |
| **1927** | Farnsworth Television patent | 95 |
| **1938** | Carlson Xerography | 84 |
| **1946** | First demonstration of magnetic tape recording | 76 |
| **1948** | Shockley transistor | 74 |
| **1953** | Book produced by offset printing | 69 |
| **1961** | Integrated circuit patented | 61 |
| **1965** | Moore's Law promulgated | 57 |
| **1972** | CT scanner | 50 |
| **1973** | First magnetic resonance image | 49 |
| **1975** | Digital camera | 47 |

## Notes

1 Commonly known as the Weimar Republic.

2 Adapted from David Mitch, in R. Floud and P. Johnson, eds., *The Cambridge Economic History of Modern Britain, Vol 1*, Cambridge University Press, 2004

3 While New England had remarkably high literacy compared to other American colonies, especially in the South, the rates only include white people. Native Americans and Black people, even those who were free, were rarely counted in statistics about education.

4 A very lively description of the history of papermaking can be found in N. A. Basbanes, *On Paper, The Everything of Its Two-Thousand-Year History*, Vintage, 2014. A useful article about the ancient Chinese papermaking is https://www. ancient.eu/article/1120/paper-in-ancient-china/ accessed 18 February 2021

5 Paperslurry.com accessed 20 March 2024

6 See Tech Talk 2.

7 A popular account of this is in Basbanes, op. cit., p63.

8 The Wendt Collection of old printed maps documents this changing world view very well https://library.princeton.edu/special-collections/publications/ envisioning-world-first-printed-maps-1472-1700

9 G. Tortella, 'Patterns of Economic Retardation and Recovery in Southwestern Europe in the Nineteenth and Twentieth Centuries', *The Economic History Review*, 47, pp1–21

10 Harvey J, Graff, *The Literacy Myth*, Routledge, 1979, is the leading proponent of this view, although many historians of literacy do not accept it. Once literacy is being achieved through widespread schooling, the social effects of the schooling are difficult to separate from the impact of literacy itself.

11 Basbanes, p87

12 A very good account of the inventions and commercial development of the Fourdrinier machines is found in Mark Kurlansky, *Paper*, Norton, 2016.

13 As late as 1975 German was required of all students receiving an accredited Bachelor of Science degree in Chemistry in the US.

14 Richard B. Kielbowicz, 'The Press, Post Office, and Flow of News in the Early Republic', *Journal of the Early Republic Vol. 3*, pp255–280, 1983

15  To read further see Archives Jean Painlevé (jeanpainleve.org) accessed 16 October 2022

16  The Law on the Freedom of the Press of 29 July 1881: a text that both guarantees and restricts freedom of expression, Fondation Descartes, https://www.fondationdescartes.org/en/2021/07/the-law-on-the-freedom-of-the-press-of-29-july-1881-a-text-that-both-guarantees-and-restricts-freedom-of-expression/accessed 3 October 2022

17  Both Hearst and Pulitzer stood for elected office. Both served in the US House of Representatives, and Hearst tried, unsuccessfully, to run for several other offices including president.

18  Timothy J. Jorgenson, https://theconversation.com/marie-curie-and-her-x-ray-vehicles-contribution-to-world-war-i-battlefield-medicine-83941 accessed 15 January 2021

19  https://spectrum.ieee.org/computing/software/carnegie-mellon-is-saving-old-software-from-oblivion accessed 15 January 2021

20  'State of the internet' (PDF), www.akamai.com. 2017 accessed 3 October 2022

# 6. BUILDINGS AND URBAN INFRASTRUCTURE

Steel, concrete and glass became the essential building materials of human civilisation from the 19th century onwards. Whoever possessed the technology for making these and using them in combinations that were both safe and economic was advantaged over those who lacked that ability. The history of these three great building materials is much older, and even millennia ago provided advantage to those who had knowledge of them, though not at the scale of the last two centuries. Until now, these have been made using processes that are very energy intensive. How the methods for making each of these critical materials evolves during the 21st century will be critical to humanity being able to meet the sustainable development principle of living within environmental limits. Nations that are at the forefront of that process evolution may establish a new competitive advantage.

What these three materials have in common is that the raw materials from which they are made – iron ore, limestone, clay and sand – are widely available and have been worked by humans since prehistory. For materials such as these, there is no single chemical composition that defines them. There are many different glasses, steels and concretes, exhibiting different properties of strength, longevity, thermal conductivity, etc. While early discoveries were serendipitous, there was certainly systematic development in many ancient civilisations, for example, glass in Babylonia, cement[1] in Rome, steel in Japan and Damascus. Many more developments occurred during the 19th century, both in composition and process for producing at the scale needed to transform the landscapes of modern cities. This continued well into the 20th century, including the growth of recycling of some of these long-lived materials.

The other similarity between the three core building materials is that the production of all of them is energy intensive. For glass and steel this involves melting of raw materials, while for cement there are also enormous quantities of carbon dioxide driven off at high temperature. Making these materials at scale requires a society with either substantial domestic energy resources or the wealth to import energy for their manufacture. Innovation is thus not just required in composition and properties, but for productivity in terms of energy required per unit of output. Making them in the 21st century requires substantial breakthroughs to meet the challenge of climate change.

Mastery of these materials, along with excellence in engineering, led to competitive advantage through radical breakthroughs in infrastructure – Ironbridge in 1779, the Brooklyn Bridge a century later. Even more profound was the advantage gained through some cities embracing the use of steel, concrete and glass for tall buildings, leading to all the benefits attendant to vibrant conurbations. Big, dense cities have advantages in productivity, energy efficiency and economies of scale that small towns can never achieve.

## CONCRETE

*It was a wise and useful provision of the ancients to transmit their thoughts to posterity by recording them in treatises, so that they should not be lost, but, being developed in succeeding generations through publication in books, should gradually attain in later times, to the highest refinement of learning. And so the ancients deserve no ordinary, but unending thanks, because they did not pass on in envious silence, but took care that their ideas of every kind should be transmitted to the future in their writings.*

*If they had not done so, we could not have known what deeds were done in Troy, nor what Thales, Democritus, Anaxagoras, Xenophanes, and the other physicists thought about nature, and what rules Socrates, Plato, Aristotle, Zeno, Epicurus, and other philosophers laid down for the conduct of human life; nor would the deeds and motives of Croesus, Alexander, Darius, and other kings have been known, unless the ancients had compiled treatises, and published them in commentaries to be had in universal remembrance with posterity.*

From the translation of *De Architectura* by Vitruvius, Book VII, Introduction.[2]

Sometime during the first century BCE Marcus Vitruvius Pollio composed a ten-volume work, *De Architectura*. In it he details all the principles of architecture, beginning with the assertion that a building must exhibit three qualities – stability, utility and beauty. This comprehensive treatise, written in Latin, is the only work on architecture remaining from antiquity. In several of the books the technology is described in detail, including machines, principles for aqueducts and materials. Brick making was perfected by the Romans, and they developed mobile kilns so that their standard sizes of Roman bricks could be made from clay deposits throughout the empire. Vitruvius warns against using lead for water pipes because of the danger of lead poisoning. Of significant importance among the materials of construction, Vitruvius provides detailed directions for preparation of lime and the making of cement.

The great volcanic eruptions of Vesuvius and Etna gave the Romans a large supply of volcanic ash (pozzolan). This they learned to crush and combine with limestone, which had been heated to drive off carbon dioxide so that it was reactive, and to which water was then added to form concrete. The Romans learned through experimentation that pozzolan produced a much harder and more durable concrete than sand. Their development anticipated the modern product, Portland cement, by two millennia.

Concrete was a crucial building material for the Roman Empire. Its use is visible today in famous buildings such as the Pantheon and in parts of the Colosseum, and it is also present, though not visible, as a lining for the great Roman aqueducts which remain as giant monuments to unsurpassed engineering superiority. Because their concrete could be used under water and withstand the harsh conditions of the seabed, it also meant that the Romans could construct piers and harbour walls at a scale greater than that of which any of their competitors were capable. Could Vitruvius or any of his contemporaries imagine that the 'stability' of their buildings, aqueducts and harbours meant that they would survive for more than 2,000 years, and still be things of beauty?

Books and articles about the success of the Roman Empire often focus on the importance of their political system, organisation of the military, and overall societal structure. All true. Crucial to the dominance that Rome achieved over a vast area of the known map of the world at the time was the quality of engineering and materials that they were able to use for building. And while the Egyptians had used various forms of concrete,[3] as had the Greeks and others before Rome, it was Vitruvius and his fellow engineers and materials scientists who invented pozzolanic concrete, perfected its production and use as a building material, and documented all their methodology.

Pozzolanic concrete certainly contributed to the competitive advantage that Rome enjoyed for five centuries. The raw materials – limestone, volcanic ash – and the process machinery of high-temperature kilns were widely available. It was the Roman engineers who systematically researched how to make this into a superior building material at scale. Here is Vitruvius' description:

*POZZOLANA*
*There is also a kind of powder which from natural causes produces astonishing results. It is found in the neighbourhood of Baiae and in the country belonging to the towns round about Mt. Vesuvius. This substance, when mixed with lime and rubble, not only lends strength to buildings of other kinds, but even when piers of it are constructed in the sea, they set hard under water. The reason for this seems to be that the soil on the slopes of the mountains in these neighbourhoods is hot and full of hot springs. This would not be so unless the mountains had beneath them huge fires of burning sulfur or alum or asphalt. So the fire and the heat of the flames, coming up hot from far within through the fissures, make the soil there light, and the tufa found there is spongy and free from moisture. Hence, when the three substances,[4] all formed on a similar principle by the force of fire, are mixed together, the water suddenly taken in makes them cohere, and the moisture quickly hardens them so that they set into a mass which neither the waves nor the force of the water can dissolve.[5]*

To understand the magnitude of the achievement, after the fall of the Roman Empire in 476 CE, the knowledge of production of pozzolanic concrete was lost. Copies of Vitruvius' book were preserved in libraries around Europe, in particular in the library of Charlemagne where copies were made and distributed to various churches and monasteries, but the significance of the more technical parts was not understood for a thousand years. During this time people could see the concrete in Roman construction and admire its properties, but were unable to reproduce it. In 1414 Vitruvius' work was 'rediscovered' by a Florentine scholar who specialised in old Latin manuscripts, and some decades later it was translated into Italian and began to circulate among Renaissance builders. Historians[6] of the Renaissance often cite its influence on Renaissance architecture across Europe, but its widespread availability also led to the revived use of high-quality concrete in buildings and other infrastructure, particularly in Italy.

What Vitruvius realised, without having access to any modern chemical knowledge, is that certain minerals, when subjected to extremely high temperatures, become very reactive. The reactivity of lime that had been heated was well known, but Vitruvius concluded, correctly, that the elevated temperatures of the volcano had performed the heating function for the pozzolanic mineral for him.

The great industrial development of the late 1700s and early 1800s meant an increase of several orders of magnitude in the need for cement. Factories, warehouses, commercial buildings, water and sewerage and major roads all needed to be built. This systemic increase in demand was a spur to innovation, and in 1824 Englishman Joseph Aspdin patented the results of experiments conducted in his kitchen, heating a mixture of limestone and clay to produce a cementitious material that reacted with water. Clay is very widespread[7] and England is particularly rich in clay deposits. The resulting product became known as Portland cement, because the material resembled a kind of building stone in use known as Portland stone.[8] Once this basic material was known to give a high-quality cement, innovation followed a course that is common to most materials developments – optimisation of the composition and ratios of feed material, redesign for efficiency of the kilns for heating, techniques for grinding the raw materials to achieve the optimum particle size, and addition of other minerals that allowed for control of the speed at which the cement hardened. Where there was little innovation was in energy use, because while the process was a great consumer of energy in the kilns it used cheap coal. This fit with British industry very well, as the country had moved to the use of coal for most industrial processes such as blacksmithing and glassmaking even before the Industrial Revolution. The improvements that were made occurred over the next 50 years after the original patent, by most standards a leisurely rate of change, characteristic of the heavy capital investment in such process equipment as kilns that requires a larger reduction in cost or increase in productivity before a company will change.

Following the invention of Portland cement, there was the consequent creation of enterprises to produce it at large scale, particularly in England (one of which was developed by Aspdin's son William, who himself made improvements to the process), Germany and a bit later in the United States. Thus, one of the key materials was in place for the creation of the great cities of Europe and the US. But it would require another material, steel, for it to find its full role in construction.

## STEEL

*There are three things extremely hard: steel, a diamond, and to know one's self.*
Benjamin Franklin

Iron is abundant in many ores, and humans have learned how to extract it through roasting into a pure form for millennia. When small, precisely controlled amounts of carbon are mixed with the molten iron, the resultant much harder metal (or more correctly alloy) is steel.[9] That carbon was the essential ingredient was not even asserted as an idea until 1786 and proven by experimentation a few decades later. Making and fashioning high-quality steel was an artisanal activity, but widespread from the Middle East to Japan, with the most well-known uses having been for warfare, especially swords. But well into the 19th century the quantities produced remained quite small, being used for such high-value items as specialised engineering tools, razors or abrasive files, but not for large-scale applications such as rails. One of the reasons for this was that iron production from ore was itself not being done at the large scale required until various developments in the second half of the century.[10] Once that bottleneck was removed, new processes for steel production, initially those developed by Bessemer and subsequently by Siemens and others, transformed steel from a niche metal to a major component of the industrialised society.

In the systems diagrams of Chapter 1, the importance of research and engineering expertise in developing the processes for converting raw materials to final products was stressed. A certain amount can be done without knowing what the basic science is, simply by carefully controlled experimentation – an empirical approach to research. This was what the Romans did with their concrete, although Vitruvius correctly surmised what it was about pozzolan that made it a desirable material. Scientific knowledge in various fields has developed at different rates over the centuries since Galileo, Boyle and others laid the foundations for modern science in the 1600s. In the second half of the 19th century, one of the areas of science that was becoming systematically understood was metallurgy, which is effectively the science of alloys. What had been a more than 3,000-year history based on trial and error plus intuition changed during this period. As seems to happen quite often in science and engineering, expertise develops in geographic centres, often, but not always, driven by a concentration of industry (discussed further in Chapter 8 on transport) in those places. One such example for metallurgy is Sheffield, in the UK, which had been a leading place for manufacture of knives and cutlery since the 1400s. In the 19th century Sheffield became a place where steel technology evolved through research and engineering, including the addition of metals such as chromium to produce the first stainless steels. It was in Sheffield that Benjamin Huntsman developed the technology for making steel by heating iron in a large crucible and adding carbon in a controlled way with thorough mixing. While many tried to imitate this process, which Huntsman kept as a secret, Britain had, from 1740 through to about 1825, a monopoly on high-quality steel, albeit still in small quantities compared to what was to come.

England's reputation in engineering and the industrial developments that were associated with it was one of the motivations for William Siemens, younger brother of Werner, who had invented and developed the electrical dynamo in Germany, to move to England in the 1850s. He is associated with many inventions and industrial improvements, but the greatest is the open-hearth furnace for steel-making. This achieved the required high temperatures by capturing combustible gases given off in the process and burning them to produce additional heat. It was because William Siemens had strong grounding in the emerging scientific understanding of heat, learned during his studies in Germany, and in metallurgical engineering gained in England, coupled with the Siemen brothers' sense of how to build an engineering business, that this major transformation of steel manufacture occurred. England and Germany were leading centres for metallurgy as a science, and this gave them leadership positions in all aspects of steel as it emerged into a dominant role at the beginning of the 20th century.

But the United States had something in addition to these strengths: the ability to mobilise large sums of capital to build enormous enterprises. These were, and to some extent still are, led by larger-than-life individuals like Andrew Carnegie, John D. Rockefeller, Cornelius Vanderbilt and Henry Clay Frick, who combined their talents with those of the giants of finance like J. P. Morgan.[11] Carnegie and Frick led the dramatic expansion of the steel industry in the United States. Moreover, these two industrialists were never so in love with the process they were using (Bessemer steel) that they would not shut it down and build new plants using better methods when they were available (open hearth steel). They were also relentless in their drive for increased productivity from the operations, usually meaning more steel made by fewer people. As discussed in Chapter 2, a passion for increased productivity is crucial to sustaining competitive advantage. Still, as the 20th century progressed, particularly in the last third of the century, other companies and countries took the leadership in steel away from the United States, and particularly from its largest company, US Steel.

## CONCRETE AND STEEL TOGETHER

A large concrete block has extremely high compressive strength, that is, a heavy weight can be put on top of it without it cracking or disintegrating in any way, but it has poor tensile strength, meaning that when it is pulled or bent it deforms or breaks comparatively easily. Concrete is thus excellent for the foundations of buildings, where the weight of the upper floors is a compressive load, but poor for beams which need to flex with variable load. This weakness was remedied, initially by placing iron bars or some sort of iron lattice within the concrete, through experimental advances in the 1850s, just as larger quantities of iron were becoming available. Concrete had only begun to be used as a building material in a significant way from about 1800, but it was this development, reinforced concrete with iron bars added, that began to expand its utility.

Once steel started to be manufactured at large scale it became the material of choice for giving concrete its required tensile strength. The mating between these two substances is nearly ideal, a result of certain physical as well as chemical properties. The thermal coefficients of the steel and concrete are almost identical, so minimal stresses are introduced due to temperature variation. The concrete flows around the steel, completely coating the surface, and if the embedded metal is either roughened or has some more complex structure such as a spiral giving it a bigger surface area, the bond between the two enhances durability in multiple directions. Finally, the chemistry of concrete (see Tech Talk 7) means that it is quite alkaline, and this leads to a reaction between the surface of the steel and alkaline components in the concrete forming a protective layer that inhibits corrosion. A perfect marriage!

Reinforced concrete and the use of steel in construction were not the results of serendipitous invention. They built on two centuries of development of the sophisticated mathematics and engineering for mechanics of solids.[12] This work, beginning in the mid-1600s, took place almost entirely in a few European centres, primarily in France, Germany, England, Switzerland and Italy. In the 20th century the centre of activity became more American, especially with the development of computers and the need for computational methods that were suitable for programming.

Because of the mathematical foundation that existed, such innovations as beams in the form of an H cross-section or stressing the steel bar while the concrete is poured around it, could be conducted and improved in a systematic way. Only those countries that valued and provided rigorous education in mathematics and engineering were able to take a leadership role in the use of these materials.

## CONCRETE AND STEEL – COMPETITIVE ADVANTAGE IN BUILDINGS

Steel in many shapes and sizes, at reasonable cost, became available in the US and Western Europe towards the end of the 19th century. Reinforced concrete technology was also in common use. Leading architects saw these materials developments as an opportunity to revolutionise what could be built. Most important was the realisation that a steel framework, riveted together, could provide an exceptionally strong yet lightweight structure for a building, and that this could be taken to heights not previously possible with concrete or stone. When concrete or stone form the skeleton of the building, the walls on the lower floors must be very thick, and windows kept small, to retain the compressive strength required. All this changed with steel, so both the height and the appearance of buildings became radically different. Tensile strength also enabled building higher, taller building needing to be able to flex in the wind. Once again sophistication in computational methods as well as testing models in wind tunnels were crucial technologies, based on mathematical analysis by the French mathematician/engineer A. J. Fresnel.

The result of these advances in materials was of course the skyscraper. Admittedly, the first one, the Home Insurance Building in Chicago, completed in 1885, was pretty far from the sky, at 10 storeys (42 metres) high, but that a building of that height could inspire awe indicates what a revolution this was. The columns of the outer structure and all the interior beams were made of steel. The exterior stone was then hung on the exterior frame, becoming known as curtain walls. This first tall steel building was designed by the Chicago architect William Le Baron Jenney.

His architectural practice became a centre of design and innovation in tall buildings, employing such future luminaries as Louis Sullivan.

At about the same time as the first tall buildings were being constructed in Chicago, Gustav Eiffel was building his tower in Paris, to serve as a symbol and centrepiece for celebration of the centenary of the French Revolution. While much taller than the Home Insurance Building (300 metres), it is not a habitable building, though platforms at several heights were used, and still are used, for restaurants or viewing. The Eiffel Tower did make use of the same computational techniques to optimise its ability to withstand wind stress, and it also incorporated some unique forms of elevators, designed by Eiffel. While it became a symbol of Paris, admired by some and derided by others in the French establishment, the engineering expertise involved in its construction was not translated into the development of the built environment. Hence, a great tourist attraction but no competitive advantage.

For a tall building to be a practical place to work requires more than a stable building. Elevators are the first requirement. Throughout this book there are examples of inventions that facilitate other advances being made just at the right time, or even earlier, with a few that occur much later. One such invention was the elevator, developed by Elisha Graves Otis in 1853 and developed into a major business by his sons after his death in 1861. Otis's key invention was not the elevator itself. Hauling a platform or cage up and down using a rope and pulley was ancient technology. What Otis invented was a safety brake, so that even if the rope or cable (which by the time of the first skyscrapers was also made of steel) was to rupture the elevator would not free fall down. It was demonstration of the safety brake which was able to persuade people that it was safe to ride to work, and later to their home, in an elevator. The development and commercial availability of electric motors (see Chapter 7) also occurred at just the right time for implementing electric elevators, by the early 1900s replacing hydraulic lifts that had been in use in factories and warehouses.

There were many other problems for the first tall buildings. If only gas for lighting (aside from natural light) had been available, gas would have had to be piped throughout the building, creating a considerable hazard.

But this was about the same time as the first availability of electric lighting, not just the bulbs themselves but all the infrastructure to bring electricity from generator to user. Heating was a challenge, and the skyscraper prompted the development of steam heat. Similarly, ventilation was much more challenging than just opening a window as in a smaller building, as there were now interior spaces far from windows. The engineering of air flow ventilation systems coupled with large new electric fans solved this problem. And no one had ever provided sanitation for a tall building – from fresh water to waste removal. This required piping and pumps.

Chicago, and then New York, became the cities for tall buildings. Manhattan had a particular advantage, being built on bedrock that could support the weight of taller buildings. Most important though was that the city regulators and planners, after some initial hesitation, embraced the idea of using the new materials to build tall structures. Indeed, they fostered competition to have the tallest building in the world. By contrast with New York and Chicago, other US cities such as Boston, Philadelphia, Los Angeles, and Washington passed laws limiting the height of buildings. Similar laws were passed in Paris and London. Thus was the stage set for competitive advantage of cities based on tall buildings.

It is not possible to get accurate measures of the economy of cities or city regions over time. The legal borders of cities change, and the extent of suburbs forming part of the city region is always varying. What is possible is to know, to some level of precision, is the population (using census data) and population change may act as a reasonable surrogate for economic development. In the 1800s New York was always dominant. It was the city with the highest population from the very first census and every one thereafter. But many other cities were more or less equal – Philadelphia, Boston, Baltimore, New Orleans, Chicago, San Francisco. They grew at different rates through the century. This changed with the skyscraper. Chicago was the fifth most populous city in the US in 1870 (a year before the fire) becoming the second most populous in 1900. During this period the population of Chicago grew three times as fast as Philadelphia and 2.6 times the rate of Boston. Even New York, with its already large population, grew at a faster rate than Philadelphia or Boston. Some of this growth related to immigration, but New York and Chicago continued to grow with the ongoing construction of taller buildings, culminating in the Chrysler Building (1930) and the Empire State Building (1931).

Land in city centres is always in short supply[13] and the only way to deal with this is to build vertically. It took time for people to adjust to working in skyscrapers, but gradually they became something like mini communities, with a diverse set of shops at the concourse level, a familiar set of faces that one saw on the elevator regularly, a corps of workers who maintained and cleaned the building, and numerous facilitators of business productivity, especially as floor sizes increased. There was also, coincident with the skyscraper, a change in the office population, which had been almost entirely male. This shift in gender balance started with the wide-spread introduction of the typewriter in the beginning of the 20th century, leading to large numbers of young women being employed. Telephone switchboards also meant more women coming into the buildings. All of this made the skyscraper an attractive place to work.

As the cost of energy increased, particularly from the 1970s onward, there was a recognition of the energy efficiency of high-density cities. New York State, despite having very cold winters, and in parts very hot summers, is ranked 50th out of 50 US states in energy use per capita, a consequence of the work done in densely populated buildings and the use of public transport that such an urban design requires.

The economics of tall buildings should primarily depend on the rent that can be achieved by adding more floors.[14] However, not everything that is built relies on rational economics. Buildings can be taller as a source of civic pride. Having the world's tallest building, a distinction held by the Empire State Building for more than 40 years, can trump rationality. Tall buildings are a tourist attraction, and derive material income from visitors to viewing platforms.

Irrationality also goes in the other direction. The City of London knew the importance of skyscrapers for a very long time before it started to build them, but had in place a law that nothing could be taller than St Paul's Cathedral, or could obstruct the view of it. Paris, despite having the second tallest structure (not a building) in the world for many decades, did not allow tall buildings in the centre, eventually creating an area 3 km outside the city limits at La Défense where taller, but still not very high,[15] buildings were permitted. Within Paris, the only tall building is Tour Montparnasse, constructed between 1969 and 1973, with 59 storeys. After it opened, Parisians were so revolted by it that a law was passed prohibiting buildings greater than seven storeys within Paris.[16]

Densification of city centres by building vertically in the 20th century was then, and remains today, a source of competitive advantage. As the world's population urbanises, in part because agricultural productivity requires far fewer people in rural areas, in part because people see cities as where opportunity is greatest, those cities that embrace tall buildings, as New York and Chicago did at the beginning of the 20th century, achieve competitive advantage. One might argue that this cannot be sustained, after all there is not much barrier to deciding to build skyscrapers in a new city. But this is clearly wrong, because once a city has become attractive to business it is hard to dislodge it, or to mobilise the capital to build somewhere else. New York, London, Singapore (another city/state that on independence embraced the skyscraper) attract businesses, talent, capital and all the ancillary benefits such as a rich cultural scene that are difficult to displace or replicate. Steel and concrete were the materials that allowed this to happen.

## GLASS

*A main object of the present invention is to improve the manufacture of flat glass in continuous ribbon form so that a better surface and flatness are obtained for the glass before annealing than have heretofore been achieved at such stage of production.*

*A further main object is to obtain a greater rate of producing sheet glass than is at present possible by the usual commercial drawing methods.*

*Another main object of the present invention is to produce flat glass in ribbon form, the faces of which have a lustre of a quality such as that known as fire finish, on emerging from the annealing stage.*

> *Still another important object of the invention is to produce by rolling*
> *methods that flat glass known as window glass, whereby the finished product*
> *is transparent and has a high-quality lustre.*
>
> GB Patent 769692, 1953, on the invention of float glass by Alistair Pilkington
> and Kenneth Bickerstaff

There are many materials crucial to construction of buildings in addition to concrete and steel, although none that are used in the same quantity. However, during the 20th century glass became increasingly important. The contrast between skyscrapers such as the Empire State, with 6,514 small windows, and later buildings of comparable height such as Sears Tower (now called Willis Tower), with 16,100 large windows, is due, in part, to the availability of large sheets of high-quality window glass from the 1960s onward. This was made possible by the invention and subsequent commercialisation of the float glass process by Pilkington Brothers in Liverpool, a process which is used for more than 90 per cent of all plate glass in the world today (Tech Talk 9).

While the British might have secured substantial competitive advantage from float glass in the 1960s (a time when the country could have badly used that sort of advantage), they did not. Pilkington chose to license the process to leading glass companies around the world. The company had spent large sums of money over a lengthy period to get the process to work and felt it needed to recoup some of that expenditure quickly.

Licensing is a model that others have used, for example for some chemical processes. It reflects two things: a sense that the lowest risk way for the company to be profitable is to teach others and collect royalties on the intellectual property the firm possesses, and a belief that the company cannot mobilise the amounts of capital required to achieve a dominant position on its own. If one were to take a positive view of this strategy, it is that in a globalised manufacturing and trading environment this is the way business will and should be done. Nonetheless, more aggressive companies do not forego their competitive advantage through a licensing model.[17]

The change from skyscrapers with small amounts of glass and progressively smaller floors as one ascended, to the glass-curtained boxes, beginning with Lever House in New York and culminating in such buildings as One World Trade Center, was also a result of a complete rethink of the steel framework by Fazur Khan at architects Skidmore, Owings & Merrill in 1963. The novel approach, known as tubular construction, takes all the load on the outer frame, though there are also designs with an interior structural tube. This reduces the amount of steel and concrete required for the building and leaves at least half the outer surface available for glass. As tubular design advanced, skyscrapers were realised in a variety of shapes, from the simplest square or rectangular cross-section to circular and more irregular and unusual shapes. All of which, as an advance in construction, would not have been useful without the invention of float glass.

Once again, the combination of advanced engineering to reconceive an established design was able to enhance the competitive advantage of those cities that had already embraced the skyscraper. The cityscapes of New York and Chicago were transformed by these new glass buildings, with spectacular views (often of other glass skyscrapers) attracting premium tenants. However, these buildings are often much less energy efficient than the earlier generation, both in cold weather, where they leak heat, and in hot weather, when considerable amounts of air conditioning are required.

The new generation of skyscrapers were seen by other cities as a way of enhancing their competitive position. Of older cities that had not embraced tall buildings before, London has been the most striking example. In the 1980s a massive construction project took shape in the area of old docks near the city centre known as Canary Wharf, attracting many large financial firms. More recently, from a very few modest-sized towers, central London now includes a growing number of distinctively shaped tall office and mixed-use buildings with nicknames such as the Gherkin, the Shard, and the Cheese Grater. Both Canary Wharf and the newer construction have enhanced London's position as one of the world's great financial centres. The London construction also illustrates the impact that these vast steel/glass developments can have on the city. Canary Wharf was built on derelict land of little value. On a normal weekday it now has more than 120,000 people at the site. Even the Shard development has rejuvenated an area at one end of London Bridge that was proximate to the more valuable areas of the City of London but was itself not very salubrious or valued. The Shard, and other new buildings, as well as infrastructure changes, have transformed this.

## THE INFRASTRUCTURE OF CITIES

It is not just buildings that make a city prosperous. A dense population needs to be able to move about efficiently. In Chapter 8 the materials used in transport will be discussed. Certainly, public transport is key to the viability of cities such as New York, London and Singapore, and the lack of efficient high-speed public transport to the poor competitive position of Beijing, Bangkok, Lagos, and many others.

Advertisement for the Roebling company, makers of the cables for the Brooklyn Bridge.[18]

The transport system, and the foresight in cities such as London, New York and more recently Singapore to build this as an enabler of growth rather than as a belated response to it, was crucial. But cities such as New York and Hong Kong, where parts of the city are separated from the centre by rivers or tidal estuaries, also depend on interconnection through bridges and tunnels to replace ferries.

The first, and arguably still the most beautiful, example of this was the Brooklyn Bridge, which opened in 1883. At that time Brooklyn and New York (comprising Manhattan and the Bronx) were separate cities. When the plans for the bridge were begun in the 1860s, suspension bridges, so common today, were not a proven technology. Indeed, some that had been attempted suffered calamitous collapses.[19] Wind tunnels and other sophisticated engineering tools, including computations other than by hand using a slide rule, were not available to test models. But well-educated engineers did know about stress/strain relationships, and how to make hypotheses regarding wind on bridges, the loads due to traffic, etc.

Universities did not embrace engineering as a suitable discipline for study and research. In 1824, when the Rensselaer School (later Rensselaer Polytechnic Institute) was founded in Troy, New York, it was the first technological university in the world. In the US this would change in the coming decades with the founding of similar institutions such as Brooklyn Polytechnic in 1854 and MIT in 1861, and the subsequent growth of engineering education in the US with universities founded by states under the Morrill Act in 1861 ('An Act Donating public lands to the several States and [Territories] which may provide colleges for the benefit of agriculture and the Mechanic arts') which became known as Land Grant universities. It was the resulting supply of well-educated engineers and an engineering research establishment to rival that of Germany which was to prove crucial to the competitive advantage of the US in infrastructure that emerged in the period after the Civil War.

One of the leading graduates of Rensselaer was Washington Roebling. His father John, a self-taught German immigrant, had built a business in New Jersey making cable with wrought iron strands bundled together, replacing rope with a material offering superior strength and durability.

When Washington joined him in the business, they saw the possibility of using the cable to provide the needed support and tension for a suspension bridge. This is just the sort of thinking characteristic of those who build great enterprises. Such bridges could use far less iron than the bridges being built at the time following the famous English Ironbridge (1781) and the later cantilevered Forth Rail Bridge in Scotland. Having successfully built several suspension bridges over smaller distances, they took on the massive project of the Brooklyn Bridge over the East River.

The emergence of steel as a mass-produced material led Washington Roebling to design and have fabricated steel wire cables, twisted and bundled to achieve the required strength. This was among the many engineering innovations of the Brooklyn Bridge that Roebling achieved, but it is the one most connected to the use of a newly available material. Because of his engineering education, Roebling and his engineering team designed the bridge to withstand six times the load to which he thought it would be subjected. In fact, because the cable manufacturers did not follow his specifications (cheated him!) what was achieved was a factor of four rather than six.

The skyscrapers rising in Manhattan coupled with the growth of residential housing where more land was available in Brooklyn (and later in the contiguous areas of Queens and further east on Long Island) transformed the Brooklyn Bridge from a convenience to what could be recognised as a source of competitive advantage for New York City. By 1900 the separate cities of New York and Brooklyn had merged, eventually including Queens, the Bronx and Staten Island with Manhattan.

One bridge was not enough, not nearly enough. The Williamsburg Bridge, also connecting Manhattan and Brooklyn, was opened in 1903 (and remained the longest suspension bridge in the world for more than 20 years), designed and initially used only to carry public transport as trolleys and railways. The Manhattan Bridge (in between the Brooklyn and Williamsburg) was opened in 1909, as was the Queensboro Bridge connecting Manhattan to Queens. Later the George Washington Bridge (1931) spanned the Hudson River to New Jersey and the Triboro Bridge (1936) connected Queens, the Bronx and Manhattan. The Bronx Whitestone Bridge was opened in

1939 and the Throgs Neck Bridge in 1961. Eventually Staten Island was connected to Brooklyn by the Verrazano Bridge in 1964. These are just the main bridges. New York City is interconnected by dozens of other bridges built from the late 1800s to the 1960s.

Tunnels were not neglected as part of this massive infrastructure development, with the Holland and Lincoln tunnels connecting Manhattan to New Jersey (1927 and 1937), the Queens Midtown Tunnel (1940), and the Brooklyn Battery Tunnel (1950). In addition to these tunnels for vehicles, between 1900 and 1933 New York built 17 railway tunnels. While a few of these were for long-distance trains, most were to allow the commuter system for both the city ('The Subway') and the suburban lines to bring workers to and from Manhattan.

Railway tunnels are less of a challenge than vehicular tunnels. The tunnelling challenge is the same, but with electrified trains there are no hazardous gases emitted in the tunnel. For a long tunnel, like the Holland Tunnel, the carbon monoxide emissions from the car exhausts would mean that the air in the tunnel would quickly render the environment hazardous to life. Prior to the Holland Tunnel only short vehicular tunnels had been built, principally in London.

The chief engineer of the tunnel, for whom it was later named, was Clifford Milburn Holland. A civil engineer educated at Harvard University, he devoted his brief working life (he died at age 41) to the design and construction of tunnels. Though Harvard, along with Princeton and Columbia, were universities from before the creation of the United States, they were slow to add science and engineering to their degree offerings. Columbia tried, unsuccessfully, to convince Rensselaer Polytechnic to merge and move to New York City. Harvard sought to incorporate MIT into the university. Again unsuccessfully. But by 1900 Harvard was offering engineering degrees and Holland graduated with a degree in Civil Engineering in 1906.

Holland had been the engineer in charge of five of New York City's railway tunnels before becoming chief engineer of the proposed vehicular tunnel under the Hudson River. Holland and his senior engineer, Ole Singstad, designed the ventilation systems, the first use of vertical shafts to remove the contaminated air and bring in fresh air. Using a scientific approach with the limited instrumentation available at the time, they refined their designs until they had a good margin of safety. This project involved collaboration with the University of Illinois, Yale University and the US Bureau of Mines, drawing on expertise from ventilation of coal mines. To evaluate the principles test tunnels were built at the campus of the University of Illinois, which had the only professor who studied ventilation in the US, and at a coal mine. The ventilation towers contain 84 fans, 42 for exhaust and 42 bringing in fresh air, so that the entire atmosphere in the tunnel is replaced every 90 seconds. Sadly, the stress of the construction project overwhelmed Holland, leading to a nervous breakdown and death from a heart attack while trying to recover his mental health at a sanatorium, just one day before the final connection between the two sides of the tunnel was to be made. His successor as chief engineer died six months later. The design and construction of the Holland Tunnel was, nonetheless, completely successful, and it became an important enabler of New York's growth and prosperity.

Steel and concrete are the key materials of the tunnels. The tubes under the river are supported by a series of large steel rings, and the pressure load of the water and rock is taken by 48 cm of concrete covering the rings. The materials were, by that time, not novel, nor was the technology for their fabrication. But the ability to combine intellectual and experimental resources from across the university and government sectors to solve problems was a key enabler. The origins of university/industry collaboration are with the chemical industry begun by Justus Von Liebig in Germany, and carried on by his most famous student, August Hoffmann, first in Britain and then back in Germany. Their work concentrated on trying to develop organic chemistry to found a pharmaceutical industry. But doing this at great scale was a US innovation in the mid-1800s. The idea that universities, at least some of them, could have as part of their mission the collaboration with industry and government on difficult problems was then, and remains, a great competitive advantage for the US. Its origins on the academic side are with the Land Grant universities, which used their expertise in agriculture (certainly not a subject that would have engaged the great universities of the Ivy League, Oxford, Cambridge, etc.), and this carried over into these same universities being highly collaborative with industries across a range of engineering problems. It formed the basis for the key contributions of many academic scientists and engineers during World War II, leading to the massive expansion of government funded university research after that. The US Bureau of Mines was also an exemplar of government taking on a role conducting research and development in support of key industries, as well as in the regulation of those industries. This continues in the US with the large laboratories of the Departments of Energy, Agriculture and Defense, and of course in many other countries as well.

## COMPETITIVE ADVANTAGE FROM BUILDINGS AND INFRASTRUCTURE – PAST AND PRESENT

Concrete, steel, glass and other materials played a major role in the development of America's great urban centres. But these materials were available to many, even within the same country. What it took to obtain massive competitive advantage from the materials was imagination, boldness and appetite for risk to build higher, even when the builders could not know if tenants would follow. The theme of entrepreneurial risk-taking to build great enterprises, recurring throughout this book, was true for urban infrastructure as well. The enterprise builders realised or at least gambled in the early days of the Great Depression in the early 1930s that better times would come, indeed, that buildings like the Chrysler and Empire State would inspire confidence in the future. The leaders of government in what became competitively advantaged city regions also acted boldly and decisively.

The technical challenges of building the early and later skyscrapers, the Brooklyn Bridge, the Holland Tunnel – these could only be solved where there was a commitment to engineering education and excellence. Countries embraced this at different rates, and even within the same country some portions of the university establishment were slow to engage with engineering as a respectable discipline. Over time, the great academic centres of engineering education became, as well, centres of engineering research, involving faculty and students in the solution of difficult industrial problems. When government grants and research laboratories were added to this mix, an ecosystem of collaboration was established. It was this entire system of education, research, and collaboration that ultimately resulted in sustained competitive advantage.

Competitive advantage required city leadership to understand the importance of infrastructure for connectivity, launching projects in the face of technical as well as financial uncertainty. During the period from 1850 to 1950 that connectivity involved railways, trolleys, bridges, tunnels, and of course electrical and telephone networks. Much of this was created as a public good. In our own time it is bandwidth and the infrastructure for communication of massive amounts of data, words and pictures. Communications infrastructure is as important today as bridges, tunnels and railways were over the previous two centuries and requires the same vision, determination, engineering expertise and recognition of how competitive advantage can be achieved. Once again it is visionary national and regional leadership that sees this new infrastructure as a public good.

## BUILDINGS AND INFRASTRUCTURE TIMELINE

|  |  | YEAR BP |
|---|---|---|
| -150 | Pozzolanic concrete use in Rome begins | 2172 |
| -100 | Vitruvius' ten-volume book *De Architectura* | 2122 |
| 125 | Completion of large concrete dome of the Pantheon in Rome | 1897 |
| 1414 | Rediscovery of pozzolanic concrete in Vitruvius' manuscript | 608 |
| 1824 | Joseph Aspdin Portland Cement patent | 198 |
| 1856 | Bessemer blast furnace for steel | 166 |
| 1865 | Siemens-Martin open hearth process for steel | 157 |
| 1867 | Monier steel-reinforced concrete | 155 |
| 1883 | Brooklyn Bridge opens | 139 |
| 1885 | Home Insurance Building completed in Chicago | 137 |
| 1889 | Carnegie Steel Corporation formed | 133 |
| 1890 | Gypsum addition to control speed of concrete setting | 132 |
| 1912 | Brearly invention of stainless steel | 110 |
| 1927 | Opening of the Holland Tunnel | 95 |
| 1931 | Empire State Building completed | 91 |
| 1948 | Durrer basic oxygen furnace for steel | 74 |
| 1953 | Pilkington float glass process | 69 |

## Notes

1 The terms cement and concrete are often used interchangeably, though this is incorrect. The minerals, usually limestone and silicates mixed into a paste with water, are cement. When these are combined with aggregate such as sand and stones the hydration reactions (see Tech Talk 7) bind the mixture into what is correctly called concrete. So, it is correct to speak of a concrete pavement, or a concrete mixer, rather than a cement pavement or cement mixer.

2 *Vitruvius, the ten books on architecture*, Vitruvius Pollio, translated by M. H. Morgan, Cambridge: Harvard University Press, 1914, https://archive.org/details/vitruviustenbook00vitr_0/mode/2up

3 Joseph Davidovits is famous for positing that the pyramids are primarily not made out of carved stone blocks but cast from something like a cement mixture, see Joseph Davidovits, *Ils ont Bâti les Pyramides: Les Prouesses Technologiques des Anciens Egyptiens*, Paris: J.-C. Godefroy, 2002, and https://en.wikipedia.org/wiki/Joseph_Davidovits

4 Slaked lime, rubble, pozzolan.

5 Chapter VI of Book 2 of *De Architectura*

6 De architectura – Wikipedia, https://en.wikipedia.org/wiki/De_architectura

7 The availability of clay of various forms and composition, and its properties of forming a hard and stable substance after heating, were well known from ceramics for food containment, as discussed in Chapter 3.

8 This has become an oft trodden path to acceptance of a new material – call it by a name that is familiar to the user so that they can believe that this is a product whose properties are familiar to them.

9 The name is derived from an ancient Germanic verb meaning to stand firm.

10 An excellent concise history of the developments of iron ore processing and steel production processes during the period 1850–1900 is given by Vaclav Smil in *Creating the Twentieth Century*, pp153–67, and will not be repeated here.

11 The modern-day equivalents being such company builders and investors as Jeff Bezos, Bill Gates, Elon Musk and Warren Buffett.

12 An excellent historical overview of the mathematical developments from the 1600s to the present can be found in 'Mechanics of solids – The general theory

of elasticity', Britannica, https://www.britannica.com/science/mechanics-of-solids/The-general-theory-of-elasticity

13  Assuming a city grows with a centre. Los Angeles and Houston showed that a more distributed model is possible for big cities, but at the expense of much larger commuting distances, with less viability for public transport.

14  'Do skyscrapers make economic sense?', USAPP, https://blogs.lse.ac.uk/usappblog/2020/08/01/do-skyscrapers-make-economic-sense accessed 14 April 2022

15  Even in 2020, after several generations of buildings at La Défense, the tallest building is 50 storeys.

16  It is a Parisian joke that the best view of Paris is from the top of Tour Montparnasse because that is the only place from which the building cannot be seen.

17  This point is made in the context of a discussion on technical innovation by James Utterback in 'The Dynamics of Innovation', Educause, https://er.educause.edu/articles/2004/1/the-dynamics-of-innovation and more extensively in Chapter 5 of his classic book *Mastering the Dynamics Of Innovation*, Harvard Business School Press, 1994.

18  John Roebling, The Brooklyn Bridge: A World Wonder, http://www.brooklynbridgeaworldwonder.com/john-roebling.html accessed 14 April 2022

# 7. ELECTRICITY

*Be it known that I, Thomas A. Edison, of Menlo Park in the State of New Jersey, United States of America, have made certain new and useful Improvements in Furnishing Light and Power from Electricity, of which the following is a specification.*

*The object of this invention is to arrange a system for the generation, supply, consumption, for either light or power, or both, of electricity, that all the operations connected therewith requiring special care, attention, or knowledge of the art shall be performed for many consumers at central stations, leaving the consumers only the work of turning off or on the supply, as may be desired – in other words, to so contrive means and methods that electricity may be supplied for consumption in a manner analogous to the systems for the supply of gas and water without requiring any greater care or technical knowledge on the part of the consumer than does the use of gas or water, in order that economy, reliability, and safety may be insured.*

Opening text of System of Electricity Distribution, US Patent No. 369,280

By the beginning of the 20th century all the technical inventions and their commercialisation were in place to construct electricity networks and do some useful things with the power they supplied. Edison, and many others, started at the device end – lighting, sound recording, motors – but it was Edison, Charles Steinmetz and Nikola Tesla above all who realised that these devices had to be part of a system that began with electricity generation and included what we would now refer to as transmission and distribution.

Everything required for electricity supply and use was possible, and electricity began to be available through small networks linked to generating stations in cities, especially in the eastern US and Western Europe. Although initially installed just to provide lighting, this was not a particularly good use of the generating capacity, both homes and offices being accustomed to using natural light during the daytime. As electric motors became available, electricity could replace various sorts of steam driven motors for industrial processes. When this happened, demand for personal, commercial and industrial use increased rapidly.

One could be tempted to say that the innovations had occurred and now it was merely a matter of rolling these out. This wrong thinking is commonly heard in respect of many innovative technological developments. In fact, just the opposite is the case. The first electric lighting offered only modest advantage over the best gas lamps – indeed the Welsbach mantle, offering a bright, steady gas light, was invented after the electric light.

What had been invented, designed and built at (modest) scale was version 1.0 of electricity supply and use. To the trained engineer and engineering business it looked not like a finished product but a set of opportunities to improve reliability and durability, increase functionality, ensure safety, while lowering cost throughout the system. One way that cost is reduced is simply through scale, as described in the systems diagram in Chapter 1. But another way cost reduction happened for electricity was through innovations in materials used for the supply and distribution of power. Materials also played a key role in device development, from the filaments of light bulbs to plastics used for insulation and structural support. In the second half of the 20th century semiconductor materials became pervasive and drove forward a complete transformation of nearly every aspect of electricity use, eventually even including the light bulb as it transitioned through fluorescent lighting, to halogen bulbs, and became the LED (light emitting diode).

Stepping back to the beginning of the electricity era, starting with transmission/distribution and continuing to devices, the materials demands fell into two main categories: conductors and insulators. Conductors can be in the form of wires, where the objective is to get as much of the electricity from source to user as possible, or filaments (for example in light bulbs) where the objective is to dissipate the electricity as light or heat. In electrical terminology the conductors can have either low or high resistance.

Insulators, by contrast, must prevent electricity flowing through a wire or device from leaving that wire and short-circuiting to either another wire or a user of the device. Improvements to both conductors and insulators were materials problems to be solved.

## ROUTES TO COMPETITIVE ADVANTAGE

But how, in the invention, initial deployment and evolution of grid electricity as a pervasive energy source in our domestic and business lives, did national competitive advantage arise? Which of the routes to competitive advantage outlined in Chapters 1 and 2 are applicable here?

The graph on the previous page gives unmistakable evidence of the correlation between electricity use and GDP (using the older, now abandoned method of GNP) in the US during the 20th century.

There are four distinct periods, characterised by 'hits' to the country's output during two world wars and the Great Depression, but the correlation with electricity output remains. In the period from 1947 to 1983 this correlation holds for each subsector of domestic (residential), commercial and industrial use.[1] Correlation is not causation. If the hypothesis is that electricity is at least a requirement for economic growth, it is necessary to understand the systemic elements that link the two. The authors of the National Academies report took exactly that approach, in which the factors correspond well to the drivers discussed in Chapters 1 and 2. The key diagram is:

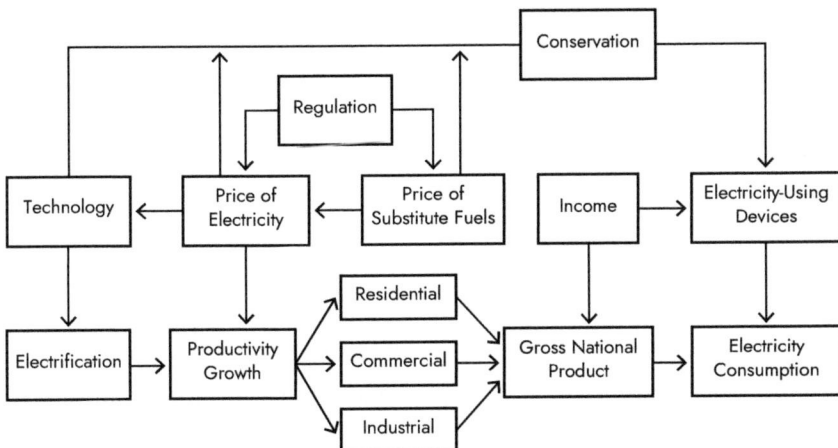

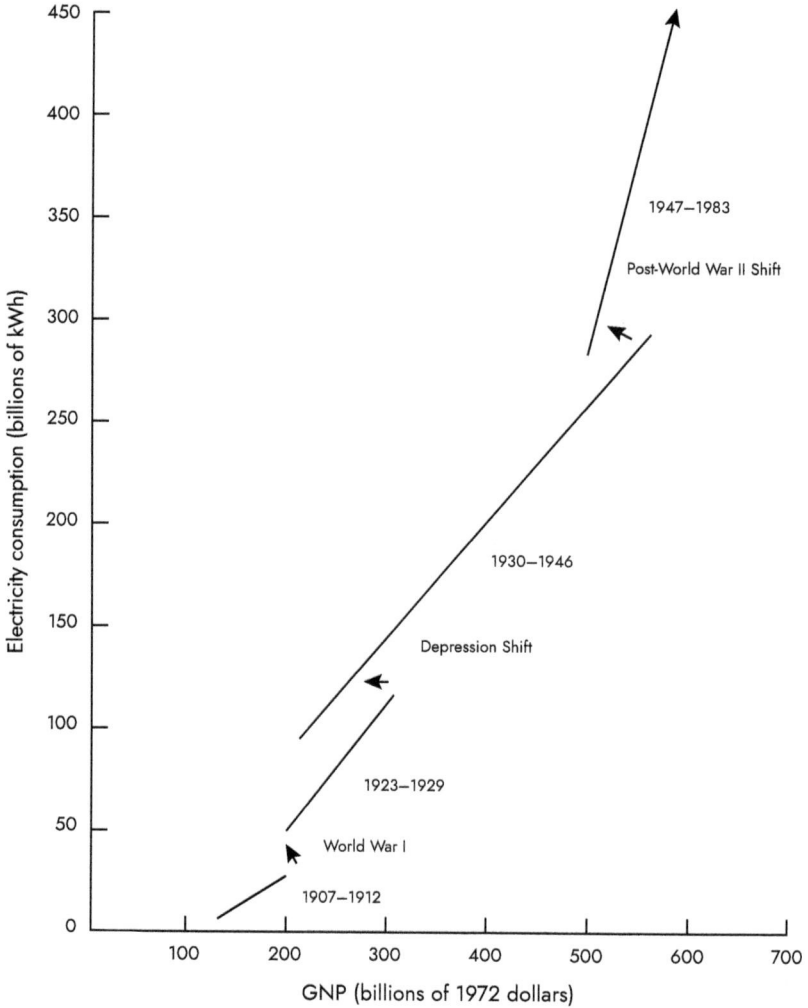

Relationship between electricity consumption and GNP in the United States during the 20th century.

This is a specific example of the systems archetypes that were shown in Chapter 1. The point to focus on here is that electrification leads to productivity growth across all sectors of society. In industrial processes, this productivity comes from safety, efficiency, low onsite pollution, and durability of electric motors when compared to steam-driven alternatives. In commercial settings, it arises from the cost and quality of lighting but also from air conditioning, and in the residential sector from both lighting and a wide array of motors incorporated into devices that free individuals from work in the house, allowing time for them to be productive in other roles in society. These pervasive gains in productivity meant, and continue to mean, that a society with access to a reliable source of electricity was/is advantaged over others that are not.

The productivity gains from electrification are thus quite different from those described in Chapter 4 on the industrial revolution in textile manufacture. There, machinery allowed production many times faster than human handiwork. The immediate impact led to violent protests and destruction of machines by the Luddites. By contrast, electricity's impact was more diverse from the outset, more generally beneficial; the existence of a grid supplying power reliably encourages many new technological developments, indeed continues to do so more than a century later.

Education is a crucial part of sustainable development. Widespread low-cost electricity has also been a differentiator in education. Even before computers became integral to education, proper lighting of schools, use of projectors, electricity for laboratory experiments, the ability for pupils to do homework and read in the evenings, built a competitive advantage for those with electricity. Computers and internet access increased this advantage by orders of magnitude.

From the viewpoint of a resident of the US or Western Europe in the 2020s, it is easy to assume that electricity as a source of competitive advantage had long since disappeared. Electricity in these countries is readily available, and reliability is generally greater than 99 per cent. But for much of the world this is not the case. Between 750 million and 1 billion people have no access to either grid or decentralised electricity. For another 1–2 billion the reliability of their electricity grid is so poor that they cannot obtain the productivity gains in any of the sectors – industrial, commercial or domestic.[2] The rural population is particularly disadvantaged in electricity access (as it was for many decades in the US and Europe). Unreliable grids, with frequent outages, are the norm in parts of South Asia and much of Africa. In many places grids are at the mercy of severe weather events that regularly leave whole regions without electricity for days or weeks.

Moreover, the quality of many grids is so poor that it has been difficult to integrate large-scale renewables without causing the reliability to deteriorate further. While perhaps 60 per cent of the global population meets Edison's goal of 'leaving the consumers only the work of turning off or on the supply, as may be desired' many billions of others are disadvantaged.

In some parts of the world the electricity grids are in State ownership, and in other parts they are private companies, though regulated by a government body. But historically, like most major infrastructure, they were either built by the State and privatised sometime later, or government policy facilitated the private company development of the grid, for example by granting an effective monopoly, or providing right-of-way for erection of towers. What is common to all grid developments is that a nation saw the importance of electrification for economic development and well-being of its citizens. Much like roads, water and sewerage, and in some places gas for heating and cooking, the grids were seen by forward-looking governments as a public good. (None of these are pure examples of a public good, because they are not available for free,[3] but their cost is low enough to be affordable by virtually all citizens and businesses, and there is sufficient supply that use by any consumer or even by a sector does not deprive others of the good.)

Electricity was also an example of another route to national competitive advantage, already mentioned for steel in Chapter 6, through the creation of huge enterprises, with their ability to raise substantial amounts of capital, their control of intellectual property along with the engineering resource to industrialise the manufacture of products. Thomas Edison's original company merged with a rival in 1892 to form General Electric (GE), which remained for more than a century one of the premier global companies in manufacturing equipment for the growth of the electricity industry. George Westinghouse formed Westinghouse Electric in 1886; it developed and acquired many key technologies, which were brought to commercial production. RCA, the Radio Corporation of America, was formed in 1919, just after World War I, in an effort by the US government to control aspects of radio transmission deemed essential for national security. It was initially the product of the acquisition of the US assets of Marconi

Telegraph by GE, instigated by senior officers of the US Navy. This was designed to thwart foreign ownership of international radio transmission infrastructure. To gain further control of intellectual property associated with this, patents and radio stations owned by Westinghouse, AT&T and even United Fruit Company were consolidated within RCA, so that these companies became substantial shareholders. On its formation, RCA was required to have only US citizens as officers of the corporation. These companies attracted some of the greatest minds behind the development of electricity, including, among others, both Edison and Westinghouse, as well as Nikola Tesla, Charles Steinmetz (sometimes called the father of electrical engineering), William Stanley and Daniel O'Conor (inventor of Formica, which was originally developed as a substitute for mica as an insulator in electrical devices, hence the name 'for mica').

Aluminium was also to prove a key component of the electrification process, and Alcoa Corporation emerged from the original producer, the Pittsburgh Reduction Company, formed in 1888 to commercialise the discovery by Hall of a route to aluminium from bauxite (see Tech Talk 10). Alcoa held a monopoly position over the patents and manufacturing technology for both producing aluminium and many of the products in which it was used for more than 50 years.

These giant corporate enterprises were central to both the electrification of the United States in the first half of the 20th century, and to bringing to market the array of products using electricity. As such, they ensured that the US would have a preeminent gain in productivity from electrification, securing competitive advantage and propelling the country past established world-leading countries, particularly those depending on empire for their wealth.

## MATERIALS AND ADVANTAGE IN ELECTRICITY

If the foregoing discussion has focused on grids, that is because grids are the dominant form by which electricity is delivered to all consumers today. Generally, these grids cover large regions, whole countries in many cases (though not the US) and even, through interconnectors, several countries. For example, Great Britain's national grid is connected to France, Holland, Norway and Ireland, with more interconnectors planned. If one was starting today, at least in certain countries, with generation of electricity from solar photovoltaics the cheapest form of electricity, minigrids or very localised generation would probably be the preferred route, rather than a large national grid. But it has been the case throughout the development of electricity in the 20th century, and into the 21st, that countries and regions without any grids or unreliable grids are severely disadvantaged.

For a grid, the key material is the wire through which the electricity is transmitted and distributed. This old photograph shows the first transmission line from Niagara Falls to the city of Buffalo. The spacing of the poles means that even for this short distance (37 km) more than 5,000 poles were required, i.e. a small forest was cut down to build this line. Why are the poles so much closer together than what is seen today?

Early electricity transmission line from Niagara Falls to Buffalo New York.

When the first grids were built, the wires carrying the electricity were copper. This was an obvious choice, copper being an excellent conductor and very abundant. Some of the biggest copper mines are in the western part of the United States, such as Bingham Canyon in Utah. But copper also has a disadvantage. It is heavy (dense), and that is why the poles need to be closely spaced to support the cable. The density also makes it difficult for workers to string the cables between poles. The solution to this problem came from what is arguably the most important materials breakthrough of the late 19th century, a route to low-cost aluminium from its main ore, bauxite. This development, described in Tech Talk 10, had a major impact on food preservation and transport, and it revolutionised the transmission of electricity.

The roots of the Hall-Héroult process for aluminium are in electricity, as the process involves passing an electric current through a molten solution of the ore with cryolite. Hall built Bunsen-Grove cells to produce his electricity, building on the work of others in the emerging field of electrochemistry that generated current from a chemical reaction occurring in the cell, and used that to conduct another chemical reaction to produce an element in its pure form.[4] Héroult received money from his mother to buy a dynamo, based on work of Michael Faraday, using it to generate DC power for his process. Electrical innovations from earlier in the 19th century thus enabled a development that would be instrumental in the expansion of electricity in the 20th century. Moreover, quite early in the commercial expansion of the business of aluminium production, the factory was moved to Niagara Falls so as to have access to large amounts of lower-cost electricity.

From the beginnings of commercialisation of the Hall process in America by the Pittsburgh Reduction Company in the late 1880s, there was a major effort to find new uses for the metal.[5] While consumer applications such as cookware were of interest, the big targets were industrial applications, particularly in buildings, transportation especially the rapidly growing automotive sector, and electrical transmission for the equally dynamic spread of electrical grids.

When the electricity industry was new, copper was readily accepted as the wire of choice. The first barrier to change was price. As mentioned above, the advantage of aluminium is its lower density, which is 30 per cent that of copper. The disadvantage is that its conductivity is only 65 per cent that of copper. These can be traded off, so that a length of thicker aluminium wire that has comparable conductivity still weighs only 47 per cent of its copper counterpart. Even without savings on installation, labour costs and poles, this simple trade-off yields a price for aluminium where it is preferred to copper for wiring.

## When trees fall on A. C. S. R. lines,
*there is less probability of line failure . . . . . .*

A.C.S.R. is more reliable, too, when short circuited by trees and other superimposed mechanical loads on the line. When arcs strike A.C.S.R, they jump around and form only surface burns. Arcing tests show that A.C.S.R. is less apt to burn deep.

Then, too, 50% to 60% of the strength of A.C.S.R. lies in its steel core, well protected from surface burns. Thus, if 1/5 of the aluminium were burned away, the loss in cable strength would not be more than 1/10. Cables of uniform material cross-sections suffer in direct proportion to the amount of surface-material burned away.

Remember these facts and what they mean to both distribution and transmission lines in greater safety and shorter outage time.

A.C.S.R., bare and insulated, is available for all types of high or low voltage distribution, transmission and other services. Bare or insulated all-aluminium cable is also offered for feeder lines, electric railways and public utility or industrial purposes.

A representative will be glad to call and discuss the advantages of A.C.S.R. for any installation you may have in mind. Please address ALUMINUM COMPANY of AMERICA; 2458 Oliver Building, PITTSBURGH, PENNSYLVANIA.

ALUMINUM CABLE STEEL REINFORCED

ALCOA ALUMINUM

*April 23, 1932 — ELECTRICAL WORLD*

Advertisement for Alcoa cable used in electricity transmission.

For longer-range transmission applications, there is also the issue of strength. Aluminium wire is not as strong as copper. Once again, a materials innovation was the answer. In 1908 an Alcoa electrical engineer devised a cable with a steel core and six aluminium strands wrapped around it. This weighed 20 per cent less than copper and was 57 per cent stronger than a copper cable with equivalent conductivity. This is illustrative of a key aspect of product competitive advantage. If you have a big advantage over a competing product, but some weaknesses, find a way to trade off some of the big advantage to compensate for the weakness. This new product, aluminium cable steel reinforced, or ACSR, rapidly gained a place in the US market, and was a major use of aluminium production by Alcoa, which was a monopoly provider. An advertisement in a trade magazine touted its advantages in case of storms.

What was clear from the outset was that despite these advantages, the ACSR cable required an array of technologies to be developed for testing, handling, erection and splicing so that workers could install it safely and provide a reliable supply. The utilities had to be educated in the merits of the new product, and they required extensive technical support. As Smith notes,[6] 'The requirements for technical support proved to be ongoing. [Alcoa created the job of] cable supervisors – engineers who assisted in the selling – who were then assigned to help customers maintain the line and to perform occasional troubleshooting. In the process, innovation became a necessary adjunct to the service-support relationship.[7] Clamps had to be devised to hold both the steel and aluminum, or to tie ACSR in with copper so as not to get electrolytic action. Such accessories became a very important and profitable part of the line.'

## COMPETITIVE DISADVANTAGE WITHIN THE US – RURAL ELECTRIFICATION

*Executive Order 7037 Establishing the Rural Electrification Administration.*

*May 11, 1935*

*By virtue of and pursuant to the authority vested in me under the Emergency Relief Appropriation Act of 1935, approved April 8, 1935*

*(Public Resolution No. 11, 74th Congress), I hereby establish an agency within the Government to be known as the 'Rural Electrification Administration,' the head thereof to be known as the Administrator.*

*I hereby prescribe the following duties and functions of the said Rural Electrification Administration to be exercised and performed by the Administrator thereof to be hereafter appointed:*

*To initiate, formulate, administer, and supervise a program of approved projects with respect to the generation, transmission, and distribution of electric energy in rural areas.*

Opening text of Franklin D. Roosevelt's Executive Order.

With ACSR and successor products from Alcoa, and with the extensive technical support that the company provided to its customers, the US and much of Western Europe grew electricity supply rapidly during the first quarter of the 20th century, albeit with substantial disruption during the World War I period. By 1932, electricity was available to all urban dwellers in the US, almost entirely through private utilities that built entire systems from generation to transmission/distribution, and on to serving retail customers. But this same system of private enterprise could not find an economic model for rural electrification, leaving 90 per cent of US farms without grid electricity. Moreover, the lack of electrical power meant that rural economies depended completely on agriculture, as no industrial or commercial enterprises would locate there. Processing of agricultural products occurred outside the areas of production.

In addition to the lack of rural electrification through much of the US, the problems in many southern states – Tennessee, Alabama, Mississippi, parts of Kentucky – were more severe. The area had widespread malaria, frequent flooding and navigational problems of rivers, soil erosion, poor farming practices and extremely low family incomes. Private companies supplied some power, but often at unaffordable prices.

When Franklin Roosevelt became president, he and his New Deal economic advisers saw that private electric utilities were supplying customers only where they could make a good return. The policy solution was to recognise electricity provision as a public good. Moreover, in the period of the Great Depression massively expanding the electricity grid could provide a boost to employment across some of the most depressed areas of the country. The first official action of the federal government pointing the way to an ambitious rural electrification programme came with the passage of the Tennessee Valley Authority (TVA) Act in May 1933. This act authorised the TVA Board to construct transmission lines to serve 'farms and small villages that are not otherwise supplied with electricity at reasonable rates.' On 11 May 1935, Roosevelt signed an Executive Order establishing the Rural Electrification Administration (REA). It was not until a year later that the Rural Electrification Act was passed. The idea of the REA was to make low-interest long-term loans to entities that would provide rural electrification in specified geographies, and to provide the expertise to allow this to be done successfully.

Within months, it became evident to REA officials that established investor-owned utilities were not interested in using federal loan funds to serve sparsely populated rural areas. But loan applications from farmer-based cooperatives poured in, and REA soon realised electric cooperatives would be the entities to make rural electrification a reality. In 1937, the REA drafted the Electric Cooperative Corporation Act, a model law that states could adopt to enable the formation and operation of not-for-profit, consumer-owned electric cooperatives.[8] These rural cooperatives would not have been capable of conducting this work themselves. The REA solved technical problems associated with the longer transmission/distribution lines of rural areas, established a standard package for each farm, trained crews to do the work and dispatched them to projects. In less than 20 years, including five years' interruption from World War II, the US went from only 10 per cent to more than 90 per cent of farms having electricity. This is another example of how standardisation can lead to competitive advantage, or to removing disadvantage. It also illustrates a role of government in creating a massive enterprise to serve the public good, when private finance will not find it attractive to do so.

At the same time, in the south-eastern US the TVA built massive hydro-electric generation, managing not only to provide electricity but to deal with other problems as well. To be sure, it was criticised then and later for its use of eminent domain, flooding of towns, environmental impacts and cost to the taxpayer. But it embodied the idea of overcoming a competitive disadvantage of a large region of the country through recognition that the use of materials could be a public good.

Large-scale electrification of the region proved an attraction for industry. As discussed in Chapter 4, the South had supplied raw cotton for textiles, first to England, and later to mills in the north. But with electricity, textile mills began to be built in the South, and this became a major source of economic growth. The availability of electricity in the TVA area, through large-scale hydropower transmitted by ACSR, came full circle as a source of advantage to the US during the war years, when the need for aluminium for aircraft and other military applications greatly increased. Alcoa, which had already begun to refine ore in the region, allowed its own power generation to be subsumed under the TVA and greatly expanded production to meet wartime needs.[9] In 1936 US production of aluminium was 112,000 tonnes, in 1940 this had increased to 206,000, and by 1943 to 920,000. This could not have happened without the government-led investment in power generation.

The pattern of electrification described for the US was broadly also followed in the populous Western European countries such as the UK, France and Germany. Initially small networks, concentrated in cities, with varying standards of supply, were constructed. For example, in Britain only 12 per cent of farms had electricity at the outbreak of war in 1939.[10] By the 1930s, most Western European governments saw the importance of more universal electricity supply using one set of technical standards, available at a reasonable retail price. Because of the damage to the systems caused by war, it was often just after 1945 that major government entities such as the British Electricity Authority (1948, which evolved into the Central Electricity Generating Board and National Grid in the UK; this built on earlier unsuccessful attempts going back to 1919), SEP in the Netherlands (1949) and Electricité de France (1946) were set up as government monopolies. Japan followed a similar pattern, with growth of grid electricity developing in parallel with the modernisation of the country during the 20th century. As has been pointed out by several authors[11] standardisation, whether by industry consensus or government imposition, is a key component of competitive advantage.

## ENTRENCHED DISADVANTAGE IN AFRICA AND ASIA –
## EMPIRE AND ITS LEGACY

*How is it that the connection between railway construction and imperial rule is such an immediate and palpable one that it is not hard to believe one can hear the locomotive pant and whistle even if the events took place a century ago? It is as if yesterday's British boys' (well, maybe all but one!) proverbial wish to drive an engine was realised by their later vicarious contact with the great railway age as they boarded the Punjab Night Mail or the Uganda Express. The link between empire and railways has long fascinated historians and gone on to beguile countless readers; think of such classics of the East African railway saga as M. F. Gill's* Permanent Way, *Charles Miller's* The Lunatic Express, *Ronald Hardy's* The Iron Snake *and Arthur Beckenham's* Wagon of Smoke, *Satow and Desmond's pictorial* Railways of the Raj, *or the Canadian epic (book and film)* The Last Spike. *Nor have empire novelists been immune from such excitement, as readers of John Masters'* Bhowani Junction *or Paul Theroux's* The Great Railway Bazaar *will quickly recall*

The History of the Nigerian Railway[12]

The availability of a key material – aluminium – in the form of steel-reinforced cable, crucially augmented by extensive technical support from the supplier who also provided other manufactured components for its deployment as transmission and distribution lines, transformed societies in the US, Western Europe and parts of Asia during the period 1910 to 1970 with setbacks from two world wars. Why didn't the same development occur elsewhere?

Most of Africa was left behind in 20th century electrification, as was India (including what is now Pakistan and Bangladesh), and this remains the case for much of that geographical area even today. How did this happen? The answer, simply, is empire. The British, French, Dutch, Portuguese and Belgians, who controlled much of Africa, large parts of Asia and the Middle East, were not averse to building infrastructure. Indeed, extensive railway systems were built throughout the regions of these colonial empires. An enormous amount of rail construction in India took place from 1850 to

1900 and the British began building the Nigerian Railway in 1898. The rail networks in these and other countries of European empires served three purposes: they made possible the movement of raw materials from the interior to ports, so that the wealth of the countries could be exploited without the difficulties and dangers of road transport. They facilitated the movement of colonial staff, again in an environment that could be done more safely and reliably than by roads. And finally, they allowed the local population to move to where they were required for work, and to return to their own villages when appropriate.

Infrastructure was important to colonial powers. But despite the clear understanding they had of the value of electrification to productivity of industrial enterprises, of the home, and for education, they did not build electricity grids in much of their empires.[13] The changes that electricity could bring to a society, well understood in Europe, were not changes that they wanted to happen in their colonies. By depriving them of an extensive and reliable electricity grid (there were certainly local grids and generators powered by diesel, for example to serve their mining interests, hence to exploit the raw material resources) they ensured that these countries were at a severe competitive disadvantage. When, in the decades after World War II, these colonies, one by one, emerged as independent nations, they started with this infrastructure disadvantage.

This competitive disadvantage largely remains the case today in many of the former colonial states, even after 60 to 70 years of independence, abating only in the last decade, especially with massive electrification in India.

While the creation of the competitive disadvantage can be traced back to deliberate policies of the occupying powers, the maintenance of this competitive disadvantage is in part due to the materials associated with transmission cable. The major aluminium alloy wire and cable suppliers to this day do not consider Africa a sufficiently interesting market, even though it needs vast amounts of their product.

| | |
|---|---|
| India | 306.2 |
| Nigeria | 82.4 |
| Bangladesh | 66.6 |
| Ethiopia | 63.9 |
| Congo | 55.9 |
| Tanzania | 38.2 |
| Kenya | 31.2 |
| Sudan | 30.9 |
| Uganda | 28.5 |
| Myanmar | 24.6 |
| Mozambique | 19.9 |
| Afghanistan | 18.5 |

(in millions)

Source: World Bank's Global Electrification Database 2012

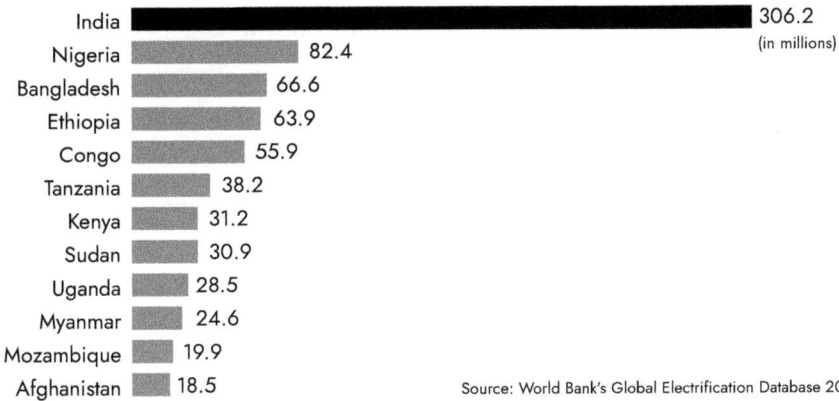

Population without electricity in 2012.

This might be due, in part, to unfamiliarity with Africa in the executive suites and boardrooms of American major players. They may be justified in their assessment, if they do one, by consulting the Ease of Doing Business rankings of the World Bank[14] which shows that of the 190 countries ranked, 33 of the bottom 60 are African. As a recent report showed[15] the suppliers have concentrated on North America, Europe and Asia Pacific. Despite Alcoa's loss of its monopoly position after World War II there are still relatively few suppliers of transmission cable. Even if a government or private entity were to have attempted a big transmission grid project in the last 60 years, they would not have had the level of technical support that is required to implement it successfully. As a result, there are some poor quality grids, highly unreliable, and vast areas with no grid electricity. Distributed solar and wind generation has the possibility of overcoming this disadvantage and is already doing so in India.

It has been noted earlier how electrification played a role in enhancing education from the earliest days, and how this role has increased in the internet age. Yet as recently as 2017 65 per cent of Nigerian schools had no electricity.[16] Many of the remaining schools get their electricity from diesel generators that may not always have a sufficient supply of fuel. Technological-based education for the modern world is thus not available in Africa's most populous country, as well as many others, and a competitive disadvantage begun 120 years ago is thereby perpetuated.

Lack of electricity also affects health. Most of the vaccines which have been so effective in eliminating certain infectious diseases around the world (smallpox, diphtheria, etc.) require refrigeration, and this has not been available across much of Africa where these diseases remain a serious problem. After lighting, refrigeration was the first widespread non-industrial use of electricity in the first half of the 20th century.

The solution to electricity access is also a manufactured materials-based answer, that is, solar photovoltaics combined with battery storage of electricity. The countries most lacking in electricity are also countries, for the most part, with high levels of insolation (the amount of sunlight during a typical year). Moreover, they have land available that is not being used for arable crops, though some of it is used by herders for grazing animals. In the first two decades of the 21st century the cost of solar photovoltaics has fallen dramatically, continuing a trend that started in the last quarter of the 20th century.

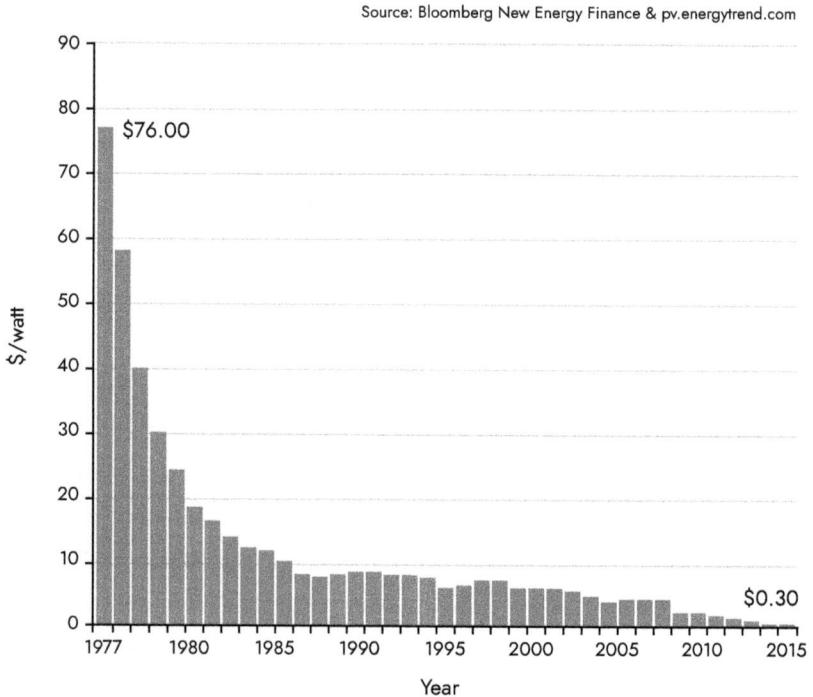

Source: Bloomberg New Energy Finance & pv.energytrend.com

Price history of silicon PV solar cells in US $/watt

This cost decrease continued, so that in 2018 it had declined to $0.18. Solar power is now the cheapest form of electricity for more than half the planet. This decrease in cost of an advanced material, crystalline silicon fabricated for efficient electrical generation, especially the dramatic changes in the past two decades, come from the most classic application of industrialisation and economies of scale that had been used for materials and devices for more than a century. They build on all the manufacturing learning from the semiconductor industry.

Previous dominant generation modes such as coal, gas, hydro and nuclear required large scale to achieve low cost, hence the configuration of big central power plants connected to users by an extensive high-voltage grid. Solar is inherently decentralised power, obviating the need for long-distance transmission. It can be deployed on individual rooftops, but larger ground-based installations, many in the tens to hundreds of megawatt range, and recently even thousands of megawatts, can be deployed to be connected by 'minigrids' for villages, towns, and even sizeable cities. The land required for this is considerable, 5 acres/megawatt, but if land is available solar materials are the answer. India has addressed its need for electricity forcefully by using solar – in 2019 and 2020 opening two solar farms each of which was larger than 2,000 megawatts, and many smaller ones.

Considering this alternative technology, effectively substituting distributed generation for transmission grid, why is it that some countries, such as India, are able to make progress on closing the gap of competitive disadvantage[17] while others, such as Nigeria, are not? As was the case with electrification in the US, Europe, and Japan a century earlier, government must see this as a public good. Where this does not happen, it must be regarded as a failure of sustainable development, as defined in Chapter 1. Sustainable development promotes good governance. Building a large solar farm, acquiring the land and importing (usually) all the key materials, then connecting this to industrial, commercial and residential users, all this requires a system of governance and a rule of law that those building the facility can rely on, because, at the very least, they must have a high degree of certainty that if they sign contracts with government entities and supply the promised electricity, they will get paid. It implies that the system is not corrupt, and that promises made by the government are fulfilled. That is what is meant by this principle of sustainable development, its end being, as stated in Chapter 1, 'building efficient economies through transparent, properly regulated markets that promote social equity alongside personal prosperity'. The materials technologies exist to eliminate competitive disadvantage in electricity access at low cost. Those countries that embrace the principles of sustainable development can do so.

## INSULATION

Well before Edison began to construct networks for supplying electricity, Samuel F. B. Morse had constructed telegraph lines. The principles of telegraphic transmission of information were extended by Bell and others to telephony. Thus, wires carrying electricity were already being deployed, albeit with much lower voltages and currents than would be used for lighting and industrial uses. The demand for materials that could provide electrical insulation, that is, non-conductors of electricity, was already present before Edison's day. Such materials prevent two wires that touch each other from forming a short circuit, allowing wires to be physically much closer to one another. They are also essential for safety.

Most early insulation of wires was by use of natural materials such as cotton thread, rubber, gutta percha,[18] glass, and later Shellac.[19] Edison did extensive experiments with materials for insulation, developing an asphalt-oil mixture for coating on wires. A particular problem when the first high-voltage cables were strung between poles was how to insulate the wires from the wooden poles, because although wood itself is not a good conductor, when it becomes wet its properties change.[20] Various sorts of insulators were fashioned from glass and porcelain for this purpose. Initially it was considered a great triumph to have glass/porcelain insulators that would work up to 10,000 volts. Today's insulators, remarkably similar in design and materials, used on high metal transmission towers, work at 500,000 volts.

Insulation for wires and cables was a manageable materials problem, though certainly one that changed completely over the 20th century as a succession of new synthetic polymers replaced natural products for wire insulation. A more difficult problem related to insulation for the components of devices. If an electrical device was to be safe, then all the components must be electrically isolated from the user. This is taken as self-evident today – the parts of a radio or television that a user can come in contact with will never be electrically 'hot'. The closest one comes in the home to electrical danger is a hair dryer, or other appliances with heaters, because the electrical element must be exposed to some extent for the device to function. Providing this insulation in the packaging of electrical devices was a major challenge in the early days of electrification of the home and workplace.

Throughout the 19th and 20th century history of materials developments for applications to clothing, transport, infrastructure and food preservation, progress in the science of natural and synthetic polymers/plastics occurring at the same time played a crucial role. Such was also the case for electrical insulation. Leo Baekeland was a Belgian chemist who immigrated to the US in 1888. He had been interested in materials and processes for the growing field of photography and set up a company in the US that developed the first high-quality photographic paper – trade name Velox. The company was sold to Eastman, giving Baekeland a substantial amount of money, but the terms of the sale included a non-compete clause that prevented him from working on anything related to photography for 20 years. His invention and commercialisation of Bakelite, a phenol-formaldehyde resin (see Tech Talk 11) made widespread availability of safe electrical devices possible, as well as having thousands of other uses. Because it was viewed as beautiful in a modern 20th century aesthetic as well as functional, Bakelite found its way into jewellery, chess pieces and pipe stems. By the time of Baekeland's death in 1944 Bakelite was used in more than 15,000 different products.

For the first two decades of Bakelite, it was manufactured only in the US and England, and these countries were able to develop electrical component and consumer product industries based around its availability. Bakelite enjoyed a significant cost advantage over other materials at this time and was a key factor in the competitive advantage through productivity gains that the US and the UK had in the first half of the 20th century.

Germany, with its excellence in chemistry, might have had similar or even greater advantage, had it not diverted so much of its chemical enterprise towards military objectives and suffered so much destruction during the world wars.

## COMPETITIVE ADVANTAGE FROM ELECTRICITY – REPRISE

With electricity there is a nearly pure case of a public good – the generation, transmission and distribution networks – established by forward-thinking governments, based on aluminium alloys, that provide all the benefits of low cost, readily available and highly reliable electricity. The sequential development of electricity showed how regions that had it (especially cities and the northern states) achieved massive economic advantage in the early part of the 20th century, and how rural electrification along with the Tennessee Valley projects brought the southern US into a more equitable position in terms of competitiveness.

The provision of electricity that was readily affordable also meant that sustainable development goals as described in detail in Chapter 1 could be achieved – children were better educated, industry was more productive, the middle class grew at a rate faster than many had thought possible. Reduction in burning liquid or solid fuels in the house or workplace for light, or diesel fuel for generating electricity, also meant that societies could live within environmental limits.

Where electricity was not available, and in many cases is still not available in a cheap, environmentally acceptable and reliable form, as is the case for much of central Africa, it means that there is an entrenched competitive disadvantage. Now technology for low-cost distributed generation is available, allowing, particularly using solar photovoltaics, for elimination of this disadvantage. It remains to be seen which governments will seize this opportunity and which will squander it.

# ELECTRICITY TIMELINE

| | | YEAR BP |
|---|---|---|
| **1749** | First use of term 'battery' by Benjamin Franklin | 273 |
| **1800** | Alessandro Volta battery demonstrated | 222 |
| **1820** | Oersted connection of electricity and magnetism | 202 |
| **1831** | Faraday demonstration of electromagnetic induction | 191 |
| **1831** | First electric motor (Thomas Davenport) | 191 |
| **1836** | Daniell cell, practical battery, used to define the volt | 186 |
| **1842** | Grove fuel cell | 180 |
| **1859** | Lead-acid battery | 163 |
| **1878** | Edison Electric Light company formed | 144 |
| **1882** | Edison generation and distribution of electricity from Pearl Street in NYC | 140 |
| **1882** | Hydropower generation, 12.5kW | 140 |
| **1886** | Hall and Heroult aluminium process | 136 |
| **1887** | Nikola Tesla establishes company to make motors and transformers | 135 |
| **1889** | NiCd battery, first rechargeable | 133 |
| **1900** | AC becomes universally accepted for electrical networks | 122 |
| **1908** | Composite aluminium electrical transmission cable | 114 |
| **1919** | Electric refrigerator, the Frigidaire. Company making it is owned by General Motors | 103 |

| | | YEAR BP |
|---|---|---|
| **1931** | Hoover Dam, 1345 MW | 91 |
| **1956** | Thyristor built at GE, enabling HVDC transmission lines | 66 |
| **1980** | Key invention for commercialisation of lithium ion batteries by John Goodenough | 42 |
| **2012** | Three Gorges Dam 22500 MW | 10 |

19 For a thorough and compelling account of the Brooklyn Bridge construction, with all the engineering, political, and personal issues involved, see David McCullough, *The Great Bridge*, Simon and Schuster, 1972.

## Notes

1   *Electricity in Economic Growth*, National Academies Press (US), 1986

2   B. J. Bulkin, I. Staffell and S. G. Mendoza, *Energy World*, April 2019

3   Most roads are an exception, as the percentage of toll roads has greatly decreased over time.

4   Henri Moissan was the first to isolate the element fluorine in pure form using electrochemistry in 1886, at the same time as Hall and Héroult were succeeding in making pure aluminium, for which he received the 1906 Nobel Prize in Chemistry, the citation of which read 'in recognition of the great services rendered by him in his investigation and isolation of the element fluorine …The whole world has admired the great experimental skill with which you have studied that savage beast among the elements.' Neither Hall nor Héroult were ever nominated for the Nobel Prize.

5   Much of the information on the emergence of aluminium cable for electrical applications comes from G. D. Smith, *From Monopoly to Competition, The Transformation of Alcoa 1888–1986*, Cambridge University Press, paperback edition 2003.

6   G. D. Smith, ibid, p91

7   The idea of lead users as innovators, put forward most forcefully by Eric von Hippel (for example in *Management Science*, 32 (7): pp791–805, 1986), was very much the case for these key developments that facilitated grid rollout and reliability.

8   A good summary of the history of rural electrification cooperatives in the US is found in https://www.electric.coop/our-organization/history/ accessed 18 June 2022

9   One of the largest plants was located in the town of Alcoa, Tennessee!

10  For a very good discussion of the process of rural electrification in Britain see www.realisingtransitionpathways.org.uk/realisingtransitionpathways/

publications/Working_papers_and_reports/RTP_Working_Paper_2015_2_
Sherry-Brennan_x_Pearson.pdf accessed 18 June 2022

11 See, for example, Daniel Immerwahr, *How to Hide an Empire*, Vintage, London, 2019, p298 ff

12 From a review of *The History of the Nigerian Railway*, https://www. britishempire.co.uk/library/historyofthenigerianrailway.htm accessed 29 November 2021

13 The situation was different in countries with extensive white settler populations, such as Australia and South Africa. Despite being part of the British Empire they followed a more European track of construction of electricity infrastructure during the first half of the 20th century.

14 World Bank. 2020. Doing Business 2020. Washington, DC: World Bank. DOI:10.1596/978–1–4648–1440–2. Licence: Creative Commons Attribution CC BY 3.0 IGO

15 https://www.tmrresearch.com/aluminum-wire-market accessed 25 May 2021

16 https://guardian.ng/features/65%-of-nigerian-schools-lack-electricity-says-un-chief/ accessed 25 May 2021

17 To be sure, India is still far from being equivalent in electricity access to a North American country; indeed, in 2019 36 per cent of Indian schools still were without electricity, https://www.hindustantimes.com/india-news/over-36-schools-in-india-without-electricity-hrd-minister/story-nEaKWjOUG02O75MOU5soiL.html accessed 23 January 2021. The Indian government claims that 99 per cent of households now have electricity, though in some places this remains very unreliable. Nonetheless, the progress in the past decade has taken place on an immense scale.

18 A latex natural product from tree of the same name.

19 A natural resin secreted by the lac bug, then dissolved in alcohol so it can be

# 8. TRANSPORT

*I will build a motor car for the great multitude ... constructed of the best materials, by the best men to be hired, after the simplest designs that modern engineering can devise ... so low in price that no man making a good salary will be unable to own one-and enjoy with his family the blessing of hours of pleasure in God's great open spaces.*

Roger Burlingame, *Henry Ford* [1]

When Henry Ford decided to produce his famous V-8 motor, he chose to build an engine with the entire eight cylinders cast in one block, instructing his engineers to produce a design for the engine. The design was placed on paper, but the engineers agreed, to a man, that it was simply impossible to cast an eight-cylinder engine block in one piece. Ford replied, 'Produce it anyway.' [2]

Of all the major sectors of the economy, transport is the one most thoroughly permeated by materials and advances in materials science. Vehicles, trains, aeroplanes, ships, bicycles – all have developed during the 19th and 20th centuries with advances in metallurgy, ceramics, polymers, composites and solid-state physics. High-strength materials are crucial to armour for vehicles and tanks. Roads and railway tracks, airport runways, and the other ancillary infrastructure of transport (including bridges and tunnels, discussed in Chapter 6) are materials stories. One should not forget rubber (natural and synthetic) for tyres, leather, plastic and fabrics for comfortable yet rugged seating, and numerous other components of passenger transport. Transport is also about solid-state electronic materials; for the 21st-century vehicle, from cars to heavy-duty trucks, the amount of computing power on board is huge and continues to grow with the move towards more autonomous vehicles.

There is an aspirational quality about personal transport, as exemplified by the conversations in Peter Menzel's book, *Material World*.[3] Families who owned a bicycle wanted to own a motorcycle, those with a scooter wanted a car, some with one car wanted a better car, a US family with two cars, a truck and a dune buggy wanted a camping trailer, and a Kuwaiti family with four cars wanted a boat. Less than 20 per cent of the world's population has ever been on an aeroplane flight, and even that percentage is after a sizeable increase this century, as hundreds of millions of Asians took their first flight.

## SECURING COMPETITIVE ADVANTAGE THROUGH TRANSPORT

For millennia, until late in the 20th century, the history of civilisation was overlaid with war, conquest and the growth and decay of empires that yielded opportunities for new empires to be built. Educational curricula teach children about the Roman Empire, Muslim conquests, the British Empire, American expansion across North America, Spanish hegemony of South America, French colonialism in Africa and Southeast Asia, the Ottoman Empire, Japanese and German military subjugation of neighbours, the Soviet Union and its control of Eastern Europe.

Many of the military and colonisation successes over these millennia have been enabled by materials. This has already been discussed in relation to food packaging/preservation in Chapter 3, and in relation to clothing (and sails) in Chapter 4. But it is all very well having shoes for your army and sails for your navy's ships if you have the roads, vehicles, and ships to move them. Likewise, Nylon played a major role for the US in World War II, replacing silk in parachutes, but the many other materials demands of aircraft were critically important to both sides in the war.

Mobile and armoured vehicles have been at the heart of 20th-century battles. The tank made its appearance during World War I in various forms, initially from the British, then the French, and finally the Americans. The Jeep was designed and prototyped in less than three months just prior to the US entering World War II. The materials requirements to achieve mobility, reliability, durability and protection from artillery attack of these vehicles, over very rugged terrain, were very demanding. In World War I the terrain was made more impassable using barbed wire, another key material, which was strung along trench lines. Tanks were the first vehicles able to move across this barrier.

The other crucial, indeed dominant, transport development affecting 20th-century warfare was powered flight. In the 21st century there is much talk about the relatively brief time from an invention to that invention having a major role in society. This is not as new a phenomenon as some might think. The Wright brothers' first sustained flight, lasting only 59 seconds, with a very fragile aircraft, was in December 1903. Yet aviation played a significant role in the war just 11 years later. Indeed, the Wright brothers immediately realised the military importance of their success and approached US, British and French governments for investment with this objective. It took several years for them to be taken seriously.

All of this is not to imply that competitive advantage through mechanised transport is only about war, rather that it was one of the things that differentiated 20th-century warfare from the millennia of armed conflicts that preceded it. Certainly from the 1800s onwards *some* of the countries implementing rail infrastructure at scale, both for passengers and freight, were able to advance their positions. The bulk of the British railway system, focused on passengers, was built during the 1840s and yielded great benefits in transforming a city/market town/village economy into a connected country. The start of the 20th century was the beginning of the great growth in road transport, both for individuals and for goods/ services. Countries that realised the importance of a paved road network to accommodate this reaped great advantage.

The first commercial airline flight took place on New Year's Day 1914. This involved both take-off and landing on water in Florida, between St Petersburg and Tampa, water being a much more forgiving surface for a pilot than concrete! Flying boats played a significant role in passenger air travel for several decades.

It is not just people that are transported. The 20th century saw a vast expansion of national and international long-distance trade of goods, both agricultural and manufactured. Materials were central to making this happen as well, especially to reducing the cost of moving goods between manufacturer and purchaser across more than one mode of transport.

## HIGHWAYS, MOTORWAYS, AUTOSTRADAS AND AUTOBAHNS – AND BEYOND ROADS

Roads are a public good,[4] and this has become even more pronounced with the general trend to eliminate tolls from most roads, or to make the amount of the toll so low that it provides funds for maintaining the road system without being a deterrent to anyone using it. The massive road building that occurred in the US and Western Europe during the 20th century is the story of two materials, concrete and asphalt; both became available at large scale at about the same time as the automobile and heavy-duty truck.

Of course, there is a reasonable argument to be made that roads are a public bad rather than a good. The materials of construction are very energy intensive. Construction of the roads requires acquiring considerable right of way, often destroying farms and dividing communities while providing little to the local economy. The vehicles using the road produce noxious and climate altering emissions, as well as noise. And the high speeds of travel result in accidents causing serious and fatal injuries. Faster road travel makes public transport through long-distance rail less attractive, removing necessary passenger revenue from the rail system.

Nonetheless, countries that invested in major road systems in the 20th century had advantaged economies as a result. Britain, France and the US were major producers of low-cost concrete, and while most of it was used for buildings, from 1913 onward concrete became a major material for intercity roads. As heavy-duty long-distance truck traffic increased, so the greater durability of concrete compared to alternatives made it a preferred material. The culmination of this in the US was the construction of the Interstate Highway System starting in 1956. As a road material concrete lasts more than twice as long between repairs as asphalt. While the grip between tyre and road in wet weather is not as good, this has been improved by a combination of roughening of the surface and advances in tyre tread design. Techniques have also been found to allow the concrete slabs to expand and contract without cracking over a wide range of ambient temperatures.

Asphalt is a mixture of various aggregates (as is concrete) but with a binder of bitumen (tar) rather than cement. Bitumen is a by-product of oil refineries, the heavier molecular weight material after the bulk of crude oil has been turned into petrol, diesel, jet fuel, etc. That the same crude oil fuelling the growth of motorised transport could also provide the paving material for the roads on which that transport would run is a lovely materials coincidence. The countries that developed oil refineries to provide fuel for transport were thus able to have a low-cost supply of road paving material available.

The battle between concrete and asphalt as the material for major highways seems never ending. But the importance of an efficient and effective road transport network, as part of a transport system for moving people and goods that includes shipping, rail and air, is not in dispute, though it is, in some countries, in transition. Of these transport modes, roads are most effective at reducing inequality in society through access to affordable mobility.[5] Set against this is the environmental impact, so that the evolution of transport in already advantaged countries is towards enabling people to move both locally and over longer distances via high-speed rail links, rather than by either road or air. Likewise, more goods may move by rail, road transport of goods may become radically more efficient in terms of fuel use, and supply chains may become shorter to reduce environmental impact. Such changes as these are already being seen in some countries. Japan, Germany, Switzerland and France have long had very punctual high-speed rail systems for passengers, and Spain has developed this more recently, leading to modal shifts to rail and away from road and air. While anyone who has been to Beijing or Shanghai in recent years will see the consequences of a city overwhelmed by cars, China has made dramatic investment in high-speed rail, leapfrogging the construction of motorways. Suzhou, formerly 1½ hours from Shanghai by car, can now be reached by rail in 30 minutes.

## Rubber for Tyres (Tires!)[6]

*Be it known that I, CHARLES GOODYEAR, of the city of New York, in the State of New York, have invented certain new and useful Improvements in the Manner of Preparing Fabrics of Caoutchouc or India-Rubber; and I do hereby declare that the following is a full and exact description thereof.*

*My principal improvement consists in the combining of sulfur and white lead with the indie-rubber, and in the submitting of the compound thus formed to the action of heat at a regulated temperature, by which combination and exposure to heat it will be so far altered in its qualities as not to become softened by the action of the solar ray or of artificial heat at a temperature below that to which it was submitted in its preparation say to a heat of 270 of Fahrenheit's scale-nor will it be injuriously affected by exposure to cold. It will also resist the action of the expressed oils, and that likewise of spirits of turpentine, or of the other essential oils at common temperatures, which oils are its usual solvents.*

Opening text of *Improvement in India Rubber Fabrics*, US Patent 3633, 15 June 1844, the process Goodyear named vulcanisation, after the Greek God Vulcan, who used fire in metalworking.

The first rubber tyres for transport were not for motorised vehicles, but for bicycles. Initially these were solid rubber, on the so-called penny farthing bicycles in the 1870s, taking advantage of Goodyear's invention of the vulcanisation process to produce a hard, resilient rubber, which was still much softer and more comfortable for the rider than a wooden or metal wheel surface. The Scot John Boyd Dunlop was the first to develop a commercially viable pneumatic tyre in 1887, although his patent was invalidated based on prior art by a fellow Scot, Robert Thomson. Not long after Dunlop founded his tyre company, Dunlop Pneumatic Tyre Co. Ltd., Michelin made improvements so that the tyre could be removed for repairs and replaced. These developments were in place and in commercial production as the practical automobile was coming into existence. The bicycle can thus be thought of as a pilot-scale test bed for evolution of tyre technology in preparation for the much more severe durability demands for the automobile. For the first five years of these new European tyre enterprises, and their counterparts in the US, bicycles and carriage wheels were the main customers. In 1901 there were 7,000 cars produced in the US, requiring 28,000 original equipment tyres, and an additional 68,000 replacement tyres. By 1918 1 million cars were made in the US (more than 4 million new tyres as spare tyres had started to become standard) and tyres were 50 per cent of the rubber industry.[7] The source of this rubber was initially Brazil and Colombia, later Southeast Asia.

Tyres as a materials story offer several lessons in obtaining and dissipating competitive advantage:

## CLUSTERS ARE IMPORTANT[8]

In about 1870 the leaders of the small north-central Ohio city of Akron, in particular Lewis Miller,[9] inventor and philanthropist, persuaded B. F. Goodrich to locate his small rubber tyre company in Akron. While there were no particular advantages to Akron, there was plenty of space and a willing workforce to be recruited, and Michigan was a centre for carriage manufacturing. As Goodrich prospered, in the run up to automobile tyre growth, Goodyear also located its plants and facilities in Akron, followed by Firestone and General Tire. Akron is about 200 miles from Detroit, where the main automotive cluster in the US had developed, and was well located to supply not only original equipment tyres but replacement tyres as well, which constituted two-thirds of the market. By 1930 Akron was within 500 miles of half the automobile registrations in the US.[10]

Clusters such as this one, and indeed the automotive cluster in Detroit, have the potential for developing a strongly advantaged position over and above what they do for the local economy. In the case of Akron, studies show that while the companies were competitors, they did not worry much about patents and intellectual property. Rather, they competed to be more efficient in production – they drove one another to higher levels of productivity, which was in everyone's interest. Though they did suffer from one big strike by the workforce in 1936, the workers also found that greater productivity was in their interest.

Innovation was in product as well as process, and each of the companies fostered a culture of innovation. Indeed, this led to many new companies, spinning out from within the original four, being set up to commercialise some of these innovations, further enhancing the size and value of the industry in Akron. A spirit of enterprise pervaded the city.

Suppliers to the industry find clusters very efficient as well, and an ecosystem of what are now usually spoken of as Tier 1 and Tier 2 suppliers grew up in and around Akron, including the nearby city of Cleveland, which also saw considerable economic growth as a consequence. This is a characteristic of all major industrial clusters, and can be seen, for example, in the oil and gas industry in Houston.

Today we think of high-tech clusters such as the Boston area, Silicon Valley, Cambridge in the UK and Lund in Sweden as growing up around major universities. Akron was the reverse of this process. The industry came first, and the university sector grew its educational programmes and research in response. The University of Akron is best known for its expertise in rubber and polymers, and what is now Case Western Reserve University in Cleveland also developed major programmes in polymer science and engineering.

Access to markets, a supportive local government, a good workforce at all levels – plant, research, engineering and management – and a network of responsive suppliers helping to solve industrial problems are all benefits of a cluster. And when the primary companies in the cluster drive each other to higher levels of innovation and productivity, they become a potent force for competitive advantage.

## TECHNOLOGY EVOLVES OVER AN EXTENDED PERIOD

By the early 1900s the tyre companies of Akron (and a few other places in the US), France (Michelin) and Britain (Dunlop) had a product that worked for cars. This product had been greatly improved by the routine addition of carbon black to the rubber, an old idea dating back to Goodyear's invention of vulcanisation, but which became widely used through the Diamond Rubber Company of Akron, which was later bought by B. F. Goodrich. Tread became standard, for better grip at the higher speeds cars were achieving. To ensure that the air pressure held, flexible inner tubes were accepted as part of the design. These had the advantage of being repairable when a nail or other sharp object punctured the tyre, essential in the early days as the roads were littered with horseshoe nails. The spare tyre was first put forward as an idea in 1904 and was adopted by some auto manufacturers in the coming decade. Before that, when a puncture (flat tyre) occurred the motorist had to disassemble the tyre, patch the tube, reassemble, and re-inflate on the roadside. For the tyre manufacturer, the spare tyre was a dream come true, selling five tyres for a four-wheeled new vehicle.[11] At this point, with productivity in the factories steadily increasing, it might have been tempting to look at the tyre as an evolved piece of technology that would not change very much over the coming automotive century. Nothing could be further from what actually happened. For strength, various methods were developed for blending strong cords with the rubber, producing bias ply tyres. Gradually at first, synthetic rubber polymers replaced some, then most of natural rubber. Radial tyres were developed in Europe and eventually supplanted bias ply tyres in the US as well, with more steel and nylon being incorporated in. The techniques for sealing tyres to rims improved so that tubeless tyres became universal. To save weight and space, smaller spare tyres were used. New technology allowed tyres that had a small leak to not collapse, the so-called 'run flat' tyre. And in the 21st century tyre pressure monitoring systems in each wheel alert the driver to low air pressure problems (still more electronic materials required!).

These are only some of the major developments. There were hundreds of smaller improvements that increased durability, comfort, convenience and safety. Almost all of these came from the research laboratories of the major manufacturers and their suppliers, especially the polymer industry suppliers of synthetic rubber. Competitive advantage can be achieved by having a superior product early in the life of an industry, and the major tyre manufacturers had that with the automotive industry. But advantage is sustained by continuous and sometimes discontinuous improvement in the technology, and only those who continue to innovate and commercialise the innovations survive. As we see with tyres, and many other industries, this is not on the time scale of a year or decade, but more than a century even for such a seemingly mundane product as the tyre.

## DON'T EXPEL YOUR BEST SCIENTISTS

Germany, Austria and Russia were the leading countries in the chemistry of polymers in the first decades of the 20th century, as they were in much of the science of chemistry, though DuPont and some other US companies began to make major advances in the 1920s and 30s.[12] In both Germany and Russia efforts were underway to make synthetic rubber, initially through the polymerisation of isoprene.[13] Even before the events in the run up to World War II, when it became clear that access to rubber would be seriously disrupted, machinations in the rubber market had led to large price fluctuations and periodic shortages of supply, providing incentives for development of alternatives. Knowledge of synthetic rubber chemistry was concentrated in a few individuals. Germany and Russia should have been able to build on these individual scientists to achieve competitive advantage in what would emerge as a key technology. But events unconnected to science changed this.

In Russia, Ivan Ostromislensky (1880–1939) was one of these key scientists. Following education and research in chemistry in Germany. There he joined the faculty at Moscow State University in 1909, and did pioneering work in synthetic rubber, and then at Bogatyr, the leading Russian rubber company from 1912 to 1917. He had many patents and publications. Ostromislensky was from a Russian noble family, however, and after the revolution life became uncomfortable for him. He moved to Latvia, but soon was invited to join the US Rubber Company in New York, working also for Goodyear and eventually Union Carbide.

Herman Mark (1895–1992) was a Viennese scientist who did much of the pioneering work in polymer science. In 1926 he began working at IG Farben, where he led the commercialisation of many important polymers, including synthetic rubbers. However, because his father was a Jew who converted to Christianity, he left his position at Farben and moved to a university post in Vienna, and then, as things became intolerable, to North America, first to Canada and then to New York, where he set up the Polymer Research Institute, founded the *Journal of Polymer Science*, and became a father to generations of polymer scientists in the US. While other great polymer scientists such as Hermann Staudinger remained in Germany, many of Staudinger's students and colleagues were forced to emigrate during the 1930s. Staudinger himself had been critical of German use of poison gas during World War I and was seen as suspect, denied funds for his research by the Nazis. What should have been massive competitive advantage for Germany in synthetic rubber was largely transferred to Britain and the US.

## DEFYING GRAVITY — MATERIALS FOR FLIGHT

*To all whom it may concern.-*

*Be it known that I, FERDINAND Graf ZEPPELIN, general-lieutenant of His Majesty the King of Würtemberg, of Stuttgart, Germany, have invented certain new and useful Improvements in and Relating to Navigable Balloons; and I do hereby declare the following to be a full, clear, and exact description of the invention, such as will enable others skilled in the art to which it appertains to make and use the same.*

F. G. Zeppelin, US Patent 621195, 14 March 1899.


*To all whom it may concern:*

*Be it known that we, ORVILLE WRIGHT and WILBUR WRIGHT, citizens of the United States, residing in the city of Dayton, county of Montgomery, and State of Ohio, have invented certain new and useful Improvements in Flying-Machines, of which the following is a specification.*

*Our invention relates to that class of flying machines in which the weight is sustained by the reactions resulting when one or more aeroplanes are moved through the air edge-wise at a small angle of incidence, either by the application of mechanical power or by the utilization of the force of gravity. The objects of our invention are to provide means for maintaining or restoring the equilibrium or lateral balance of the apparatus, to provide means for guiding the machine both vertically and horizontally, and to provide a structure combining lightness, strength, convenience of construction, and certain other advantages which will hereinafter appear.*

Wilbur Wright and Orville Wright, US Patent 821393, 22 May 1906 (but originally filed in 1903).

*To all whom it may concern:*

*Be it known that I, Hugo Junkers, a subject of the Emperor of Germany, residing at Aachen, Germany, have invented certain new and useful Improvements in Flying Machines, of which the following is a specification.*

*My invention relates to improvements in flying machines particularly of the aeroplane type.*

H. Junkers, US Patent 1114364, 20 October 1914.

The popular mythology of the Wright brothers was that they were two bicycle mechanics from Dayton, Ohio, who tinkered with machinery and were lucky enough to have the first powered flight. Nothing could be further from the truth. The brothers studied everything they could find about the aerodynamics of flight, requesting materials from the Smithsonian Institution in Washington, making observations about birds, experimenting with gliders, before setting out on an attempt to achieve human controlled, sustained powered flight.[14]

Their systematic approach led them to calculate that the engine they would need for their first plane would need to produce at least eight horsepower and weigh no more than 200 pounds (91 kg). They investigated all the automobile engines available in America at that time (ca. 1902) and none met the requirement, so they knew that they would have to build their own. They discussed this problem, particularly how to reduce weight, with an acquaintance in Dayton at the Buckeye Iron and Brass Works, who pointed them to the possibility of using aluminium, suggesting that they contact the Pittsburgh Reduction Company, the company set up in 1888 to commercialise Hall's process for making aluminium, located some 250 miles away. After some discussions and further experimentation to achieve strength as well as limit weight, they settled on making the engine block out of a 92% aluminium–8% copper alloy. In Germany considerable research was in progress on aluminium alloys designed to achieve various properties. Daimler and Benz were already experimenting with making engine blocks for automobiles using these alloys while the Wright brothers were building their engine. The engine that the Wrights finally built exceeded their requirements, developing 12 horsepower with a weight of 180 pounds.

The Wright brothers were not alone in seeing the importance of the dramatically lower cost of aluminium for flight. Zeppelin made extensive use of it in his development of 'airships', which replaced the simple inflatable balloon with a rigid frame to which a fabric cover was attached. For the frame, he used the German developed aluminium-copper alloy called Duralumin, which also contained small amounts of manganese and magnesium. This material achieved greater strength than the simpler aluminium-copper alloy of the Wright brothers and becomes stronger after it is produced (spontaneous age-hardening). This alloy is to aluminium as steel is to iron; it is that important.

If aircraft were to develop beyond the early Wright brothers' designs, the wood and canvas portions of the plane would have to be replaced by something stronger and more durable, and that meant a light yet strong metallic material. It was clear from Zeppelin's work on airships that Duralumin, or related aluminium alloys, were the only suitable answer.

It was Hugo Junkers (1859–1935) in Germany, with his company Junkers Flugzeug and Motorenwerke AG, who conceived and realised the idea of an all-metal aircraft, using a combination of corrugated iron and aluminium alloys to achieve his goal with the Junkers J1 in 1915 and an all-metal passenger aircraft in 1919, the Junkers F 13. Junkers and Zeppelin built substantial enterprises of importance to the German economy.

Duralumin was developed and patented in Germany by Alfred Wilm, but neither he nor any of the contemporary metallurgists understood how it developed its high strength. The tools to enable this understanding at the atomic level were not yet available, and too much of metallurgical research was being done on trial and error rather than a systematic basis. It was pressure from the US Navy in the years leading up to the US entry into World War I that eventually led to a theory by US government scientists for the mechanism by which the heat treatment, rapid cooling and then aging at room temperature caused alloys of aluminium with copper, magnesium and manganese to develop very high strength.[15]

At the Pittsburgh Reduction Company, and its successor Alcoa, all the effort had been on commercialising Hall's process at scale, improving reliability of production, putting in place the logistics to supply large quantities of bauxite ore, building plants near low cost sources of electricity (the development of this electricity-based industry was taking place at the same time as the electrical generation industry itself was scaling up), and lowering costs to enable more applications. In this they were highly successful. The first Hall process aluminium in 1889 sold for $4.08 per pound (50 per cent of the lowest price before that). By 1893 this was down to $.78 and by 1897 to $.36. It would fall to below $.20 before World War I. The systems thinking behind this relentless focus on increasing scale and lowering cost (while maintaining/enhancing profitability) was sound. Many of the new markets for their product was replacing some other material, and every cost reduction they achieved would open a new replacement use. However, it meant that a culture of incremental improvement, problem solving and process engineering was dominant, and that did not leave room for a research effort aimed at fundamental understanding behind what was essentially a metallurgical business. In the period to 1920 the main experimental effort was devoted to manufacturing and testing, rather than research into new products.

There was a small research effort by Charles Martin Hall himself, probably designed to keep him busy away from the main work of building the company. Hall's main interest was in electrochemistry, and he resisted the idea of creating a strong technical organisation for the company. It was only the pressure from Germany and its alloy development that made Alcoa realise that its product line needed to go beyond aluminium itself, and led, in the immediate post-war period, to the creation of a proper laboratory for research and development. The death of Hall in December 1914 removed that barrier to creating a proper technical organisation for the company. Alcoa hired Francis Frary, a research chemist from the University of Minnesota, to create this centre, which had a prodigious set of achievements over the next 20 years. For example, Alcoa had been producing aluminium of 97.75 per cent purity, sufficient for most applications. Frary's team did fundamental work on the process and developed a way to increase the purity to 99.99 per cent! This increase in purity is more than just a 'nice to have'. The nature of aluminium alloys is that a relatively small amount of another metal added to the aluminium has a big effect on properties. Choose which metal, and with experience and fundamental metallurgical understanding, one can decide what properties to optimise. But when the starting material is only 97.75 per cent pure, the properties will be affected by the 2.25 per cent impurity. With much higher purity metal, research into many more novel alloys became possible, so that by the late 1930s 40 of the 47 most important alloys in use worldwide had been developed by the Alcoa laboratory.[16]

Once again, the lesson described above for tyres held true for aluminium. Developing a working process to produce a useful product at an attractive price is just the beginning of building an industry, not the end. Competitive advantage for the US, because of Alcoa, was maintained and enhanced, but a long-term commitment to research and development of new products, understanding of fundamentals, integration of research with business and customer requirements, and above all nurturing of talent was required to do so. This commitment had to be sustained during good times and bad – Alcoa started its research laboratory in the boom times after World War I but did not abandon it during the difficult depression years of the 1930s. As a result, when the military requirements of World War II arose, the US was competitively advantaged because of its deep knowledge and commercialisation of aluminium alloys. In the post-World War II years, this translated into an explosive growth in aluminium consumer products in many sectors that continued for many decades. This is the same story as Nylon, a product of the DuPont research laboratories in the same period.

## REDUCING ENERGY IN TRANSPORT DRIVES MATERIALS CHANGE

From the beginnings of the dominance of oil as the fuel for transport in the early 20th century, global oil consumption and global GDP increased at the same rate, and this continued until the early 1970s. The post-war growth in global GDP was accompanied in the US, Japan and Western

Europe by a doubling in oil consumption over the next 25 years. The 1973 Yom Kippur War and the ensuing Arab oil embargo changed that. The quadrupling of the price of crude oil in 1974 was a shock to the economies of the countries where oil consumption was profligate.

There were several possible responses. The traditional one, not taken in this case, was a military option to capture the oil fields of the Middle East and bring them under western control. A second option was to find and bring on stream alternative supplies, in the US and elsewhere. Fortuitously oil in the North Sea had been discovered in 1969 and production began in 1975. While this did not have a measurable effect on price, it did at least improve security of supply for Europe. The third option was to use less – energy efficiency. This was the most successful of all strategies. From 1979–1993 there was no growth in oil demand, or production, but there was a 40 per cent growth in GDP. For the first time in the 20th century, energy consumption, primarily for transport but also for heating and power generation in some places, was decoupled from economic growth.

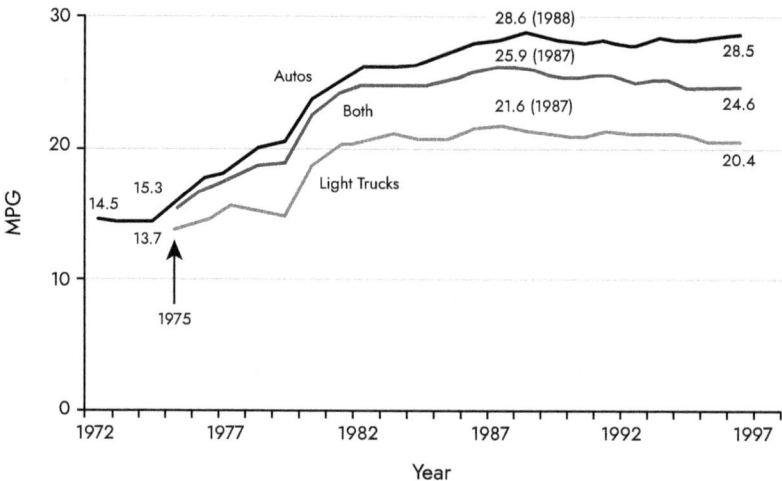

The doubling of average fuel economy of the US car and light truck fleet in approximately 10 years.

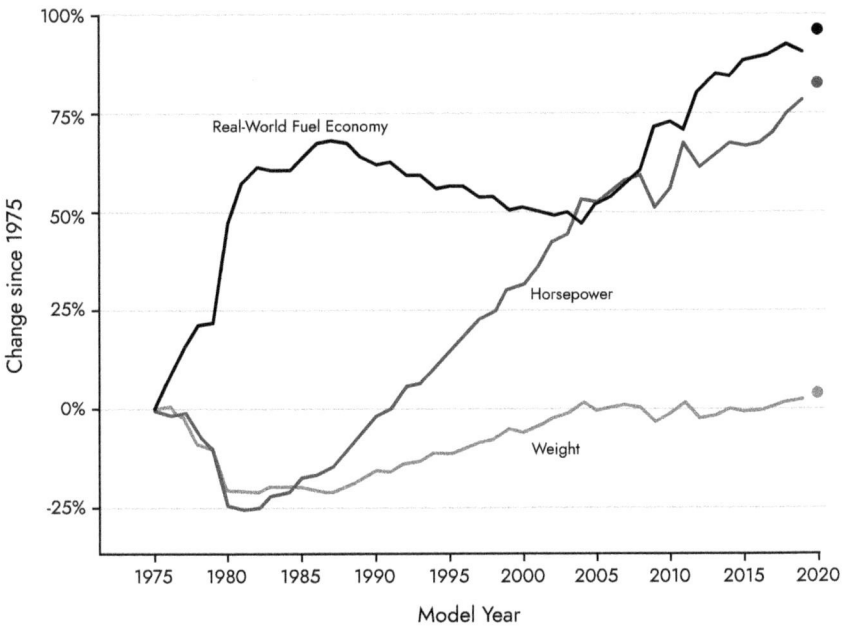

Fuel economy increased, while weight and horsepower decreased in the early years of CAFE standards in the US. In the period 1985–2000, in the absence of new regulation, weight and horsepower increased, and fuel economy declined. In the period 2000–2020 it was possible to increase horsepower, improve fuel economy and keep weight constant. Note that the numbers are different from the previous figure as this graph shows 'real world fuel economy', a different measure than the official test results used in the previous one.

This decoupling was largely achieved by a dramatic increase in the fuel economy of the vehicle fleet. It was not an industry initiative, but the result of a legislative requirement, the Corporate Average Fuel Economy (CAFE) standard introduced in the US in 1975. After the industry met these standards (which they did two years ahead of what was mandated), government failed to introduce new standards for more than two decades, and fuel economy did not improve, even though technology was available to do so.

The second graph[17] shows how this was achieved and gives the longer time perspective. The quick gain in fuel economy was done by rapid weight reduction – cars were put on a serious materials diet. A lot of metal was replaced by plastic and composite materials. The amount of aluminium in cars increased fourfold between 1975 and 2010. Steel manufacturers developed higher strength steels that could achieve the same performance at lower weight. The average weight of cars over time, even considering the gradual increase from 1985 to 2020, is even more dramatic when seen in the context of better safety, higher fuel economy and higher horsepower.

Materials innovations achieved much of the gain in the 20th century, and computer control of engines and emissions did the rest in the first decades of the 21st century. Nonetheless, persistently finding ways to replace steel by aluminium or composite materials continues to be the simplest way to achieve weight reduction and fuel economy. The average American car contained 154 kg of Al in 2010 (up from 38 kg in 1975) and this increased to 211 kg in 2020.[18]

This story, taking place over 50 years (and similarly in Europe with, on average, smaller cars and higher fuel economy) has sometimes been represented as dematerialisation, though it looks that way more from the perspective of the steel industry than the aluminium and composites business. It is, however, best described as a substitution of one material by another rather than a dematerialisation. Overall, the planet benefits, taking a full lifecycle analysis approach, and if this is the motivation for achieving the change then it is consistent with what will motivate dematerialisation in other sectors.

## MOVING GOODS — CONTAINER SHIPS

Malcolm McLean's family did not have enough money to send him to university, but there was enough to buy him a truck, from which he was expected to earn a living. Together with his brothers they built up a trucking business, hauling goods from maker to buyer around the United States. Eventually, he had sufficient funds to sell his part of that business and buy a shipping line, so he could deliver freight by sea. As he watched goods being loaded and unloaded, usually from truck to ship or vice versa, taking a long time and at great cost for the labour of dock workers, he wondered about a better way. He initially thought about just driving the truck onto and off the ship, but soon realised that this was an inefficient use of both space and trucks. In 1956 he pioneered the idea of putting all the goods destined to a single customer into an 8-foot by 8-foot metal container, which could be taken from truck to ship using a crane and unloaded in a similar way. He bought some surplus World War II ships from the government, had some metal containers manufactured, and sent a first load from New York to Galveston, Texas. Containerised shipping was born.

McLean patented this idea and built up a company that eventually was known as Sea-Land, which he later sold to R. J. Reynolds, the tobacco company, with whom he had had a relationship as deliverer of their cigarettes.

Containerised shipping was, much to the fury of the unions representing dock workers, one of the greatest advances in productivity of the 20th century. The labour required to load/unload a ship dropped from 20 to 22 men to 1 to 2. The cost, in 1956 dollars, went from $5.86/tonne to $.16/tonne, a 36-fold reduction. And the time reduced from hours to minutes.

The material that enabled containerised shipping had been invented a few decades earlier, by United States Steel Corporation. It was a particular stainless steel known as Corten (sometimes written as Cor-Ten), that contains copper and chromium. The key property of this steel compared to other stainless steels is that it is very corrosion resistant. When exposed even to moist air for extended periods of time, it forms an oxide layer that coats the material, much as aluminium does, but now with the tensile strength of steel. It is also weldable. Thus strong boxes can be built which do not degrade during shipping and can be transferred from truck or railcar to ship and back again many times.

Realising the competitive advantage from containerised shipping required two other things that are repeated themes in this book. The first was standardisation. If containers were to be widely used, they needed to be of a standard size, or a few sizes. McLean realised this, and there were long negotiations, only settled in the late 1960s, with the International Standards Organisation (ISO), which involved McLean making some of his patents available on a royalty-free basis. The second was government regulation, in this case the US government pulling back from some of its regulations on interstate commerce that required special permission for any mixed shipment across state lines. Sometimes regulation facilitates advances, and sometimes it is deregulation that is required.

At its outset, containerised shipping provided competitive advantage to the US, allowing goods to move at much lower cost between different US ports. A half century later, it made Chinese manufactured goods available at low cost in distant ports both in the US and Europe, thus transferring the competitive advantage achieved in the late 1950s. Today, more than 90 per cent of non-bulk goods are shipped by containers. Containers conceal their contents and are able to be securely locked. Technology enables detection of any attempt to tamper with the lock before it reaches its destination. As a result, theft of goods in transport has been greatly reduced.

## COMPETITIVE ADVANTAGE FROM MATERIALS FOR TRANSPORT – LOOKING BACK AND LOOKING AHEAD

The mechanisation of transport, starting with the great railway construction in the mid-19th century, and accelerating into the 20th century with both road and air developing rapidly at about the same time, has led to a huge change in how people work, win wars, or take vacation, in those countries that were able to take full advantage. Materials played a vital role in this, including steel for trains and rails, various metallic products for early automobiles, concrete and asphalt for roads, rubber for tyres, and even in the first flight, aluminium for aircraft and cars, glass for everything. Throughout the 20th century, aluminium has played an increasing role, but so have plastics, composites and electronic materials.

As with many of the other examples discussed in this book, it was not the countries with the most iron ore, oil, bauxite or rubber plantations that gained the advantage through mechanised transport. It was, as always, the technology to convert these raw materials to products, and to manufacture them at huge scale.

Transport is yet another example of the importance of clusters, for example in Akron and Detroit in the US, in establishing an industry and its supply ecosystem, access to local markets coupled with global ambitions. These clusters are not just efficient, they lead to innovation as well, so that nations that have established leadership can maintain it over decades.

Within the mechanised transport industries, workers were able to achieve middle-class levels of affluence. True, this sometimes involved serious battles between owners and unions, but with a long view these struggles can be seen as the system working its way to what was clearly in the interests of both parties. These large numbers of workers provided positive feedback to the system – not only in being able to afford new cars or take an occasional aeroplane trip themselves, but to the education of their children to higher standards, leading to more highly skilled employees for the next generation, even as less assembly line labour was required.

The US and several European countries including the UK, France, Germany, Italy and Sweden all secured these advantages from transport in the first half of the 20th century, though in Europe some of it was certainly dissipated by war. Post 1945, Japan and Korea also saw the importance of mechanised transport for attaining a more advantaged economy. These Asian competitors did not just copy the North American example, they were innovative in their production models, and in how they saw quality processes in manufacturing as central to success – a particular example of productivity that led to reductions in cost and waste.

The flaw in this beautiful story? The successful countries did not see the need for mechanised transport to conform to the sustainable development principle of *living within environmental limits*. Air quality impacts of all aspects of transport were ignored, certainly in places where the entire economy was built around personal cars rather than mass transport, until the air became unbreathable. It took a long time for epidemiological studies of tailpipe emissions, including lead, to get the attention of governments that could begin to regulate these. The technology to address these problems was certainly available, but manufacturers never showed any interest in being proactive. There was a finger-pointing culture between vehicle manufacturers and fuel suppliers, each saying the other one had to solve the problem, when it was always obvious that to radically reduce emissions required them to optimise the entire system. To this day, some political leaders in advantaged countries see agreements to reduce emissions as a bad thing in terms of competitive position. That these arguments can be made shows a lack of comprehension of the importance of living according to the principles of sustainable development as being a source of durable competitive advantage of nations.

## TRANSPORT TIMELINE

| | | YEAR BP |
|---|---|---|
| 1825 | Stephenson invents first steam engine for trains, called Locomotion No 1 | 197 |
| 1827 | John McAdam demonstrates macadamised roads | 195 |
| 1866 | Otto engine, beginnings of the internal combustion engine | 156 |
| 1870 | J. D. Rockefeller founds Standard Oil Company | 152 |
| 1871 | B. F. Goodrich relocates his tyre company to Akron Ohio | 151 |
| 1881 | First electric streetcar line | 141 |
| 1892 | Rudolf Diesel invents the diesel engine | 130 |
| 1902 | Edgar Hooley produces first Tarmac road surfacing | 120 |
| 1903 | Wright Brothers powered flight | 119 |
| 1908 | Model T Ford introduced | 114 |
| 1913 | First concrete highway | 109 |
| 2003 | Tesla, Inc. electric car company founded | 19 |

applied as a coating.

20 The cables themselves are generally not coated with insulation, because air is a very good insulator.

## Notes

1  Roger Burlingame, *Henry Ford*, A. A Knopf, 1970. It is believed that Henry Ford first said this in 1906.

2  https://www.thehenryford.org/collections-and-research/digital-resources/ popular-topics/henry-ford-quotes/ accessed 7 April 2024

3  Peter Menzel, *Material World – A Global Family Portrait*, Sierra Club Books, 1994

4  See Chapter 1 for a definition of public goods.

5  The broader subject of how innovations affect inequality in a society or between societies has been discussed by E. M. Rogers in *Diffusion of Innovations*, 3rd Ed, The Free Press, 1983, pp391–413.

6  An excellent example of Britain and America being two countries divided by a common language. The origin of the word tire as used in vehicles is obscure (attire the wheel?, or something connected to the French verb tirer?) but historically it has always been spelled with an i in the US and Canada and a y in the UK and most other English-speaking countries.

7  William Pretzer in 'How tire is made – material, history, used, processing, parts, components, composition, steps, product', https://www.madehow.com/ Volume-1/Tire.html accessed 15 May 2022

8  Porter writes extensively about the importance of clusters for national competitiveness in Michael E. Porter, *The Competitive Advantage of Nations*, 2nd ed. Macmillan, 1998, however he does not mention tyres as an example. Aspects of this work have already been discussed in Chapter 2 of this book.

9  Miller's daughter Mina married Thomas Edison.

10 An excellent analysis of the Akron tyre cluster, discussing various ideas about why it came about and its importance, is in Guido Buenstorf and Steven Klepper, 'Heritage and Agglomeration: The Akron Tire Cluster Revisited', Paper # 508 in the series Papers on Economics and Evolution, Max Planck Institute on Economics, Jena. 2009. retrievable as https://hdl.handle. net/10419/31812

11 From 1941 the US prohibited the equipping of cars with spare tyres because of rubber shortages caused by the war.

12 Cf Tech Talk 1

13 See Tech Talk 12

14 For an excellent description of the early work of Wilbur and Orville Wright on aerodynamics see David McCullough, *The Wright Brothers*, Simon and Schuster, 2015.

15 The US-German competition on aluminium alloys is discussed in George David Smith, *From Monopoly to Competition*, Cambridge University Press, 1988, pp127–131.

16 Ibid, pp163–176

17 'Highlights of the Automotive Trends Report', The EPA Automotive Trends Report, US EPA, https://www.epa.gov/automotive-trends/highlights-automotive-trends-report accessed 12 May 2021

# 9. THE FUTURE MATERIALS ENTERPRISE

In 1990 Michael Porter, looking to the future of national competitive advantage, wrote, 'What of the future? The central economic concern of every nation should be the capacity of its economy to upgrade so that firms achieve more sophisticated competitive advantages and higher productivity. Only in this way can there be a rising standard of living and economic prosperity.'[1] Yet a decade later Howard and Elisabeth Odum wrote:

> As the global crescendo of information and investments rushes toward the culmination of civilisation, most of the six billion people on Earth are oblivious to the turndown ahead. It's time for people to recognise what is happening and how they will be forced by circumstances to adapt to the future ... Like a giant train, the world economy is slowly cresting its trip up the mountain of growth. It may be ready soon for its long trip down to a more sustainable lower level.[2]

This view, that growth as we knew it in the 19th and 20th centuries is not forever, has been developed with specific reference to materials in more recent publications such as Andrew McAfee's *More from Less*,[3] in Allwood and Cullen's *Sustainable Materials*,[4] and with large quantities of factual data in Vaclav Smil's *Making the Modern World – Materials and Dematerialisation*.[5] One of the first to look at these issues of materials was Tim Jackson,[6] whose more recent books on prosperity without growth and life after capitalism are very provocative.

Having spent all of this book on historical examples of competitive advantage through materials, the look ahead to the future has three distinct aspects:

First, is national competitive advantage even possible anymore? Certainly, some thinkers feel that in the current era of global communications, extensive trade, rapid development of intellectual property, and the far greater importance of services over manufactured materials to the economy, the sorts of competitive advantage through materials discussed in Chapters 3 to 8, are no longer possible. Connected to this question is another: are any of the fundamentals that were important as laid out in Chapters 1 and 2 and demonstrated with examples in Chapters 3 to 8 no longer important – for example, education/literacy, protection of intellectual property, a growing middle class, progress towards equality and advancement on merit rather than birth or connections.

Second, if the future is about doing more with less, indeed incorporates a trend towards dematerialisation of our society, does this put an end to national competitive advantage from materials or open new opportunities of a completely different sort? To what extent does the ecological impact of human activity through materials change the game of competitive advantage? Which aspects of the principle of sustainable development from Chapter 1, 'living within environmental limits', are local, which are global?

Third, what are a few of the emerging, and in some cases already emerged, materials-based technologies that could be differentiators in the future for those nations or regions that see the opportunities where others do not? This then loops back to the first point: what is it about these materials, and these nations, which might lead to technology being exploited to achieve competitive advantage even in a globalised economy?

## GLOBALISATION AND NATIONALISM

From about 1995 to 2015 there was a gradually expanding view that competitive advantage for nations, as distinct from advantaged companies, based around such things as manufactured materials, was no longer possible in the way it had been for centuries before. This was premised on the idea that technologies moved rapidly from one place to another in the globalised economy, that the techniques for advantaged manufacturing were easily copied, even that skilled labour was becoming less important with the rise of readily available digital design tools, computer controls and automation of all sorts. Indeed, the same computer tools, such as MATLAB, produced by MathWorks Inc., are used in engineering universities all over the world to educate engineers, and then by companies that employ these engineers.

Michael Porter's 1990 book, *The Competitive Advantage of Nations*, referred to extensively in Chapter 2, had developed a theory of how nations can compete successfully, and had been influential. He dealt with the competitive rise of countries such as Korea and Japan, the steady and renewing position of Switzerland, and the decline of Britain. However, less than a decade later many began to question whether the sorts of examples he documented were still relevant in what was seen as the inevitably globalised economy.

The attack on the idea of national competitiveness was kicked off with an influential article by Paul Krugman with the provocative title 'Competitiveness: A Dangerous Obsession'.[7] This was, to some extent, a response to initiatives during President Reagan's tenure to improve the US competitive position versus Japan and Germany, who were seen to have taken a lead. It was this initiative that had led to the formation, in 1986, of the Council on Competitiveness in the US. Indeed, this council still exists, and, to quote from its current website:

*In a time of ever accelerating technological change and business transformation, the Council on Competitiveness shapes policies and runs programs to jump-start productivity and grow America's economy. The Council's membership is diverse and nonpartisan, representing the major sectors of the economy – CEOs, university presidents, labour leaders and national lab directors.*

*Together, we work to ensure U.S. prosperity by advancing a pro-growth policy agenda in Washington, DC, and then we take action ourselves convening program initiatives across the country aimed at creating public-private partnerships where new technologies are born. These member-led initiatives are shaped around the areas of innovation frontiers, CTO policy advocacy, advanced computing, and energy and manufacturing.*

*The roots of the Council trace back to 1986 during the Reagan-era Commission on Industrial Competitiveness, chaired by Hewlett-Packard CEO John Young. At the end of the Commission's work, Young created the private sector Council on Competitiveness. At the time, U.S. competitiveness was being challenged by the rise of international competition from countries such as Japan and Germany. While the country names may have changed, today's threats to our economy have only grown through globalisation. Today's competition is a race to see who will innovate and develop key technologies in artificial intelligence (AI), the Internet of Things (IoT) and 3D printing, to name a few. Future prosperity and an increased standard of living is in the balance.*[8]

Krugman's challenge to this has been summarised as:

- It is misleading and wrong to see parallels in terms of competitiveness between a nation and a company.

- Despite the fact that firms compete with each other to get a greater market share, and the success of one business means the failure of another, the success of a country or a region creates more than destroys the opportunities for others; trade among nations is not a game 'without result'.

- If competitiveness has any meaning, then it is just another way to describe productivity. The development of a national standard of living is determined primarily by the rate of productivity growth.[9] This is the area of commonality between all views on national competitiveness in the 21st century. However, it raises the question of whether a drive for continuously improved productivity is consistent with growth in accordance with the principles of sustainable development, i.e., does it meet the challenges laid down by Jackson, the Odums, and Smil?

Krugman's reaction to the idea of national competitiveness morphed into something quite different in the course of the next two decades, as the globalised economy view became dominant. An emerging view of some economists, and indeed some European countries, was that they could not compete with Asia on manufactured materials, that there was no point even trying, and that they would be best doing things like financial services or making television serials instead. Some would argue that this is not vastly different from what David Ricardo had said about trade and national competitiveness several centuries earlier. This is not to say a country gives up on being competitive. Rather it asks, why keep trying to make your auto (or chemical, electronics, plastics, steel, cement...) industry more productive than the Chinese (or Vietnamese, Malaysian, Indian…), when you can invent such new businesses as Netflix, Facebook/Instagram, Apple, Amazon, Google? Where such industries require manufacturing, just get someone else to do it for you. Hence the label on some Apple products: designed in California, manufactured in China. James Fallows, in a series of influential articles,[10] challenged Americans worried about Japan's rise as an economic power by wondering: why be unhappy if there were workers willing to work long hours for not that much money in order to produce high-quality cars and electronics that Americans could enjoy at low prices? He was later to revise his views, as did Krugman.

Globalisation also challenges the centuries-old idea stressed in this book of the importance, for such things as manufactured materials, of intellectual property as a determinant of competitiveness. For example, there is no debate over the role that the Chinese have played in reducing the cost of solar panels through mass manufacturing. This has led to revolutionary change in the production of electricity throughout the world, and China certainly has the leading role in this business. Still, many other countries are playing a key role in manufactured products for aspects of the solar industry, including the US, Germany, Israel and India. While some of this is production of components such as inverters, or design tools for large solar farms, there are panel manufacturers in India at the same scale as some of the leaders in China. There are interlocking supply chains, movement of expertise to lower-cost areas, gradations of quality and cost. None of this looks like the way a country or region-dominated industry manifested itself a century earlier. The entire industry has developed with little of importance in patented or licensed intellectual property. Solar-generated electricity is just one example of the effects of globalisation for manufactured materials. Electronics of all sorts, many large-scale polymer products, the components for producing, containing and storing pharmaceuticals, telecommunications technology, batteries of all sizes, self-driving cars – in all such industries tied to manufactured materials there has been a change that makes most of the earlier history look positively quaint.

This view, however dominant it was in the early part of the 21st century, may have been naive. Nationalism and a desire for competitive advantage of nations, specifically in manufactured goods, was not at all finished as an idea. It has had an easy transition from the US obsession in the last quarter of the 20th century with Japan and Germany to an obsession with China in the first quarter of the 21st century. It should have also included, especially in aspects of manufacturing, an obsession with South Korea, whose GDP per capita was 13.6 per cent of Japan's in 1970, 32 per cent in 2000, but 79 per cent in 2019. It has clearly been possible for countries to change their competitive position and achieve dominant global roles in specific industries during the 21st century.

European countries are not exempt from the ongoing idea of national competitiveness either. The UK, France, Germany, the Nordic countries – all are striving to find materials products where they can have sustainable competitive advantage, even if they lack the muscle of the US and China. To be sure, certain national leaders have asserted these positions on competitiveness for political advantage rather than with a clear industrial strategy to achieve them. Even those with what appears to be a determined strategy, such as China, have not always seen the limitations of their positions very clearly.[11] In this regard, the resources described in Chapter 2, knowledge, capital and infrastructure, remain highly relevant. Despite the globalisation of knowledge and supply chains, there is ample evidence that national competitive advantage in materials remains important and possible. It just may look completely different from anything that has been seen before.

A specific point needs to be made about infrastructure. The 20th-century competitive advantage from materials in infrastructure was about asphalt, concrete, steel and glass. By the end of that century, it was clear that tall buildings, tunnels, bridges and highways could be built anywhere. In 1950 17 of the world's 20 tallest buildings were in New York City. By 2000 only eight of the 20 were in the United States, four in New York. The critical infrastructure component for competitive advantage in the first quarter of the 21st century is that which is required for high-speed communications, and in materials terms that is optical fibre.[12] This is strongly differentiated by country. In 2019 (Figure 9.1) Singapore, Japan and South Korea each had 100 per cent of their premises connected fully with optical fibre for communications, whereas the UK had 12 per cent and Germany 11 per cent. It is true that the laggards are increasing rapidly, so one might say that this is not a sustainable competitive differentiator, though once business is attracted to or lost to a country it is hard to change these decisions absent a new factor.

FTTH/B leaders with 95%–99% coverage

→

■ % HP/HS 2012
■ % HP/HS 2019

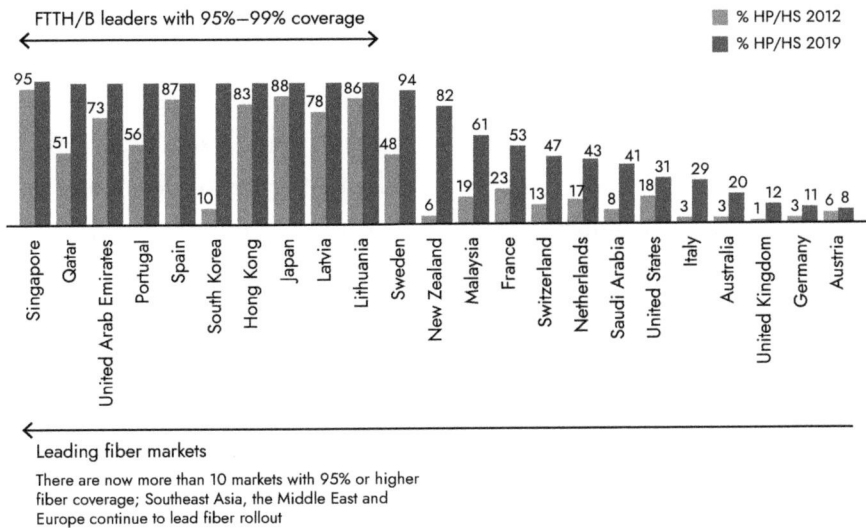

Leading fiber markets

There are now more than 10 markets with 95% or higher fiber coverage; Southeast Asia, the Middle East and Europe continue to lead fiber rollout

Optical fibre available to the home in different countries in 2019 vs 2012, showing the strong differentiation achieved by some countries.

There is a counterfactual approach that also proves the durability of national competitive advantage in the 21st century, remaining consistent with the principles discussed in Chapter 2. If there was no longer a possibility of competitive advantage for nations, then there would be no reason why some poor nations remain poor. Yet they do. Most of central Africa has remained extremely disadvantaged, while the competitive advantage of China, Vietnam, India, Malaysia and others grew. Yet countries such as Nigeria have had access to the same globalised technology and supply chain. Indeed, Nigeria appears to have some of the factor advantages, though it has a relatively deprived educational system, with a literacy rate of just 62 per cent (compared, for example, with Zimbabwe at 91 per cent), and even this is an 11-percentage-point improvement in the past decade. After a sluggish economic start to the century, Nigeria had steady growth for 10 years but has fallen back so that its total output in 2021 is similar to a decade earlier.

However, overlaid on this is that from 2000 to 2020 the population grew by 68.5 per cent. Nigeria is 136 out of 180 countries in GDP per capita in 2019.[13] Contrast this situation with Vietnam, where the GDP per capita increased by nearly 700 per cent in the first two decades of the 21st century. Looking across a large number of African countries it is clear that it is very possible for a nation that is poor to remain poor, or to lose competitive position in this globalised economy. As pointed out in Chapter 7, lack of electricity infrastructure, a heritage of empire, plays a significant role in maintaining disadvantage. There is no technical barrier to overcoming this, though the lack of a proper system of governance according to the principles of sustainable development means that the corporations that can mobilise resources and bring technical support, for example for installation of key electrical transmission and distribution, are not attracted to the country. As asserted in Chapter 1, living according to the principles of sustainable development can be a route to competitive advantage, and the counterfactual argument also shows itself here.

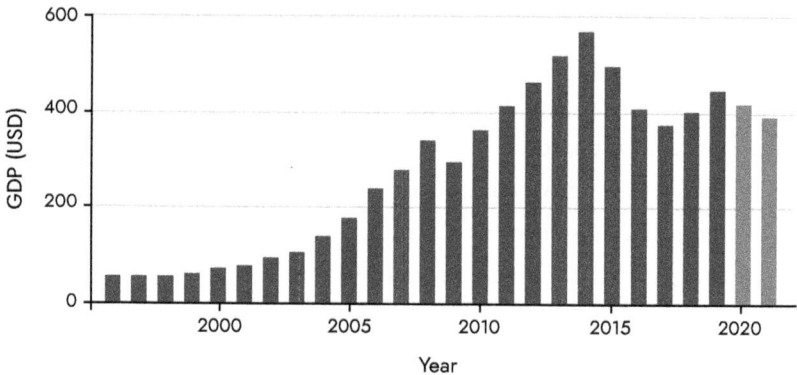

Source: tradingeconomics.com & World Bank

GDP of Nigeria — 25-year view, in USD.

## DE-MATERIALISATION: THREAT, OPPORTUNITY OR FANTASY?

Reducing the amount of material used in any application is always good business, and has always been recognised as such by astute business leaders, if not by most governments that use crude measures of output to measure success. Beginning in the last quarter of the 20th century, and accelerating in the 21st century, reducing material content has moved from being 'a good thing to do' to 'essential for the future of our civilisation'. It has become a movement, and governments have scrambled to show leadership. It is difficult for government thinking to fit the idea of reduced use of materials into their goals and metrics for national competitiveness. The headline 'GDP increased by xx per cent in the last quarter' as the bearer of good news on the economy from the government is inconsistent with reducing the use of materials in everything we manufacture.

In Chapter 8 the role of materials in reducing the weight of automobiles during the last quarter of the 20th century was discussed. Most of this was substitution (aluminium and high-strength composites for steel), but a portion also occurred by manufacturers looking critically at every component and asking whether it needed to be there in the quantity currently used, given the imperative to reduce weight and increase fuel economy while maintaining safety and reliability.

When aluminium cans were introduced for beer in the early 1960s, they weighed a similar amount to the tin-plated steel cans they were replacing. A modern aluminium can weighs only an eighth of that. Moreover, it is probably made of recycled material, with approximately 80 per cent of aluminium cans being recycled. Such is the scale of the drinks industry that making cans one gram lighter saves 200,000 tonnes of aluminium a year. With lighter cans there are consequent energy savings in transport, and for chilled drinks in cooling (because less metal must be cooled).

Dematerialisation takes several forms: the most definitive is when a product or part has been 'overdesigned' and manufacturers learn that they can get the same performance with lower material use; it can be the substitution of one material for another, allowing less to be used overall; it is, in our time, a more radical substitution such as the use of digital means for storage of information rather than paper, assuming this actually results in a net decrease in material and does not lead to information being stored both electronically and as hard copy; and it can be waste reduction in the manufacturing process, because even where waste material is reprocessed it still involves extra costs, be they labour or energy or both. In principle all forms of dematerialisation in our manufactured goods are an increase in productivity, and increasing productivity is crucial to competitiveness.

There are valid reasons to be sceptical about the drive towards dematerialisation to really deliver lower absolute consumption. Make something cheaper by using less material and demand will increase. Indeed, this is what many manufacturers believe when they drive towards lower materials consumption, and it is part of the positive feedback described in Chapter 1. Many examples of this have been discussed by Smil,[14] who has analysed most of the assertions of reduced materials use, both relative and absolute, and found that many of them do not lead to absolute reduction. Nonetheless, it is worth looking not so much to what has been done in the past, but to future opportunities.

## PACKAGING – MATERIALS CONSUMPTION OUT OF CONTROL

A more clear-cut area with opportunity for future dematerialisation is packaging. If there has been one big materials trend in society over the past half century it has been the relentless increase in packaging. With this has come vast quantities of waste, some of it recycled, but much of it not recyclable because the packaging was not designed with that in mind. It certainly began with food and the transition to self-service food shopping from markets and grocers. Some of this was for hygiene, some of it convenience in having food that could be sold by weight or volume without actually needing to be weighed or measured in the store. For the supermarket owner, it reduced waste, particularly with produce, but also with goods such as cereals and cereal products. It also offered an opportunity to influence customer behaviour, learning to select items for purchase by appearance only, rather than by smell or touch. In more recent decades, the trend to packaging more and more of the products meant that a customer did not have to decide how many tomatoes to buy. The supermarket had done that for them by packaging them in a box of four.

Even the parts of this that make sense in terms of food waste reduction, efficiency of the customer experience or supermarket profits have been carried to extremes beyond what seems sensible or consistent with 'living within environmental limits'. If lettuce is sold by the head, there is no question of it needing to be weighed or measured. To cut it up for serving is a job of perhaps one minute. Why then, alongside lettuce sold by the head, does every supermarket have a refrigerated cabinet with chopped-up lettuce sealed in a plastic bag, probably washed with chlorinated water, and possibly packed under a special atmosphere, in packaging that is not recyclable or reusable? Similarly, what improvement is conveyed to the customer experience with a cucumber by buying one shrink-wrapped in polyethylene?

Simple things can cause big problems. Labels pasted on to polyester containers must be attached so that information is available to the consumer, both before and after purchase. But these labels can have a severely detrimental effect on recycling of the packaging. But a change in adhesive to one which switches off in the caustic recycling bath, dissolving the adhesive and letting the label float away, is the simple solution to this problem.[15]

Just as supermarkets educated customers towards acceptance of packaging, there is a compelling case for dematerialisation here. In some countries it has already begun with reusable bags replacing disposable plastic or paper bags. In 2015 in the UK, a five-pence charge was introduced for a plastic bag in supermarkets and other large retailers. In just three years this small charge led to a reduction in more than a billion plastic bags being given away, a 95 per cent reduction in an area where plastic use had been growing by as much as 20 per cent per year. This is proper dematerialisation, even when one nets off the reusable bags. In supermarkets, this will continue with the simplest things like fruit and vegetables, and then, for those who lead the way, increased use of refillable containers for products like shampoo and liquid soap.

Food packaging is just the beginning, and in some ways it is the most defensible use of excessive packaging. The hardware or DIY store of today is replete with excessive packaging, of tools, light bulbs, screws, paint brushes. That every screwdriver is packed with a cardboard backing and plastic seal, often multilayer materials that do not lend themselves to recycling, is responsible for a lot of material consumption that neither enhances the efficiency of the store nor the consumer experience nor is it required to protect the product from damage. Most of this packaging thus winds up in landfill.

## REDUCING WASTE IN MANUFACTURING AND CONSTRUCTION

Waste occurs in almost every materials process used in manufacturing. Even with the most advanced aluminium beverage can manufacturing, 12–14 per cent of the metal is wasted in the process. While this is reprocessed rather than thrown away, it means additional cost in energy and labour which carries forward through the process.[16] With ca. 200 billion aluminium beverage cans each year, and the number increasing, this is an opportunity with scale! This same generic wasted material, what is left over after cutting a shape from sheet or block, pervades manufacturing. This process is called subtractive manufacturing, an umbrella term for various machining and material removal processes that start with solid blocks, bars, rods of plastic, metal or other materials that are shaped by removing material through cutting, boring, drilling and grinding. Additive manufacturing, discussed further below, is the answer for achieving radical reductions in this waste.

There has been a massive increase in construction and demolition waste created over the last 30 years in the United States. In 1990, 135 million tonnes of construction and demolition debris by weight were created and that rose to 600 million tonnes in 2018. This is a 300 per cent increase. Of the 600 million tonnes of waste created due to construction and demolition, 143 million tonnes of it resides in landfills.[17] This means that about 76 per cent of waste is now retained and repurposed in the industry, but there is still more waste being exported to landfills than the entire amount of waste created in 1990. Europe has been more successful in reducing waste to landfill through taxation, but it is not clear how much of this is due to more efficient construction and how much is waste being dumped illegally or exported out of the European Union.[18] Even with landfill tax, in the UK 32 per cent of landfill waste comes from the construction and demolition of buildings. It is estimated that 13 per cent of products delivered to construction sites are sent directly to land-fill without having been used.[19] Reduction of construction (as opposed to demolition) waste is quite amenable to progress through the introduction of new software tools at all stages of the process.[20]

## REDUCING WASTE THROUGH LONGER LIVED PRODUCTS

It is not just unrecyclable packaging, construction and demolition waste that winds up in landfill. There is also all the stuff we just throw away because it does not 'work' anymore. In terms of quantity of materials, especially with high embedded energy, the biggest culprits are appliances such as clothes washers and dryers. This is not about appliances that break down because of a fault in some part. Such faults can be reduced but not eliminated. It is about those failures being difficult or impossible to repair economically, leading to the best economic decision being replacement of the appliance. It is a well-documented complaint of consumers that manu-facturers' guarantees for such items tend to be one or two years, although they claim lifetimes of around eight years.

Electronics waste may be smaller in weight per unit, but the quantity of laptops, phones and other gadgets has become enormous. Moreover, the value of the embedded materials, and the complexity of separating them for recycling or reuse, as well as shorter product lifecycles, makes this a large materials waste problem.

Dealing with this is a question of design. When engineers build a factory, they assume that repairs will be required. They design for long lifetime and for reparability. It is curious then that engineers do not exer-cise the same design mentality for the products of the factory, for example white goods and electronics found in our homes. As a result, they get thrown away. Dematerialisation will occur through waste reduction when consumers demand longer guarantees, published ratings on expected life-times from manufacturers (like energy ratings that have had a big influence on consumer behaviour) and a well-publicised network of trained and trusted repairers of appliances.

At the other end of the spectrum in terms of quantity of material per unit are clothing items. Many of these have sacrificed durability for cost, and this has been in response to consumer preference. Examples would be t-shirts and women's pantyhose/tights. At the extreme, some clothing items are worn less than 10 times before being discarded because of materials failures. As in the case of packaging, consumers need to be wooed towards durability.[21]

## DEMATERIALISATION AND LOWER WASTE MEAN HIGHER PRODUCTIVITY

*Too much and for too long, we seemed to have surrendered personal excellence and community values in the mere accumulation of material things. Our Gross National Product, now, is over $800 billion dollars a year, but that Gross National Product – if we judge the United States of America by that – that Gross National Product counts air pollution and cigarette advertising, and ambulances to clear our highways of carnage.*

*It counts special locks for our doors and the jails for the people who break them. It counts the destruction of the redwood and the loss of our natural wonder in chaotic sprawl.*

*It counts napalm and counts nuclear warheads and armored cars for the police to fight the riots in our cities. It counts Whitman's rifle and Speck's knife, and the television programs which glorify violence in order to sell toys to our children.*

*Yet the gross national product does not allow for the health of our children, the quality of their education or the joy of their play. It does not include the beauty of our poetry or the strength of our marriages, the intelligence of our public debate or the integrity of our public officials.*

*It measures neither our wit nor our courage, neither our wisdom nor our learning, neither our compassion nor our devotion to our country, it measures everything in short, except that which makes life worthwhile.*
Robert F. Kennedy[22]

Only a few of the possible targets for reduced material use have been discussed so far. Allwood and Cullen have devoted a whole book to this, with many examples of how cement, steel, aluminium and many other materials can be used, with technologies that are already commercial, to make all sorts of products and structures more efficiently with less. Their analysis also treats the energy flows in manufacturing materials from ores and how processes can be redesigned to reduce energy requirements.

The problem to be solved is partly one of design, and a commitment to design for durability, reparability and recycle/reuse. It is, as has been made clear from the few examples discussed, certainly also one of waste and society's attitude towards that. As the world commits to tackling climate change, it is a problem of treating the embedded energy of all our products as something to be minimised rather than 'just another cost'.

All of these are problems that can be understood, then incorporated into the economy of materials manufacture and use. Still, there is a conundrum of economic thinking to be overcome. On a very regular basis, governments provide figures on GDP growth. Growth of GDP, the total of goods and services produced by the nation, is taken as a measure of economic health, which is expected to translate into personal welfare, growth in the value of investments and national competitiveness. A nation with faster-growing GDP is to be envied, one with slower or negative growth to be pitied. Yet, as the quote introducing this section indicates, this is at best an imperfect measure, and one that has been the subject of numerous more recent critiques, especially from Professor Tim Jackson.

If society sets itself on a course to use less material to achieve the same functionality, if products are designed to last longer, and if there is less material wasted, this course is one that is generally a negative contributor to GDP, even after one nets out jobs created through recycling of waste. Indeed, data indicates that for a range of European countries the quantity of waste increases linearly with GDP.[23] Similar relationships have been shown for a variety of Asian countries.[24]

The contradiction emerging here follows from what has already been discussed in Chapter 2 on competitiveness, namely that the most essential thing is productivity; to be sure this has been mostly about how things are made (labour) in the past, but it is also very much about what is made. Dematerialisation, waste reduction and durability are in this respect all gains in productivity. This is, in part, what Tim Jackson has argued for more than two decades.[25] At its most basic, we achieve productivity gains by decoupling the amount of material used or wasted from GDP. Ideally this is 'absolute decoupling', that is, GDP continues to increase but materials consumption stays constant or decreases. Even 'relative decoupling', in which materials consumption increases at a slower rate than GDP is, however, valuable.

If decoupling can be achieved in this way, then a core principle of Chapter 2, that productivity always wins in terms of national competitiveness, still holds. Nations that follow this path will be more successful than those dealing with massive quantities of waste and throwaway products. If they can, at the same time as committing to standards leading to dematerialisation, become the leaders in developing and manufacturing the products that meet these standards, then these nations will be the future leaders in the world economy. Still, thus far there has not been a demonstration that decoupling as described here is possible except in limited areas.

There is one other strand to this argument that bridges to the next section of future materials. While Jackson and others[26] have made the case against dedicating the political/economic system to growth, an alternative view, perhaps a more obvious one, is that the route to prosperity and well-being is through 'green growth', that is, our society requires a whole range of new products (materials) to transform itself into one which 'lives within environmental limits', and these products will provide the growth we have always associated with successful capitalistic society. This appeals to those who optimistically see technology as always finding a way to transition a society from one stage to another, a view supported by much of what has been discussed in earlier chapters of this book. It also encourages governments to intervene in the promotion of the green growth industries, while allowing them to continue to measure success using the same GDP per capita yardstick with which they are comfortable. A nice summary of the conflict and contradictions between the green growth and degrowth schools is in an article by Geoff Mann.[27]

## CLIMATE CHANGE AND MATERIALS ADVANTAGE

Responding to the challenges of climate change – less $CO_2$ and less methane in our atmosphere – requires a complex set of technology and behavioural changes. Many of these centre on key materials, and those nations that position themselves to secure the ability to make those materials will (in some cases already) reap competitive advantage in the 21st century. Later in this chapter the technological challenge for concrete, steel and glass is discussed. The energy intensive processes for making these materials must be radically reshaped if atmospheric $CO_2$ is to be reduced. What of new materials?

In Tech Talk 6 the process for making ultrapure silicon is described. This was developed to supply computer chip manufacture, where purity of at least 99.9999999 per cent is required. More than 90 per cent of all solar cells are made from high-purity silicon, though it needs only (!) be 99.9999 per cent pure. Until the rapid growth in deployment of solar photovoltaics (PV) began in the early 21st century, the industry relied on rejected material from computer chip manufacture for its supply. This limited the total amount of new photovoltaic installation in the world to about 100MW per year.

Silicon minerals (mainly quartzite) are widely distributed around the world. There is no competitive advantage from having a lot of raw material. Early in the 21st century, China began to invest heavily in building large-scale plants to produce silicon for PV. China's share of polysilicon production (for both solar PV and computer chips) increased from near zero in 2005 to more than 70 per cent in 2020. It is estimated that China also produces more than 80 per cent of solar panels from this silicon.[28]

The process used is not unique or proprietary. The dominance of China was achieved by application of the systems principles in Chapter 1, particularly the cost reduction that can be achieved by building an enterprise at vast scale. It used some of the capital generated from its other strong export of materials in the previous two decades to invest in new giant factories. It also allowed, indeed encouraged, many companies to be set up and compete with one another across the value chain. By producing the panels as well as the high-purity silicon it was able to maximise capture of the value in the final product.

Is this a sustainable competitive advantage? Probably not. India, where solar PV is being installed at the rate of about 16 GW per year, is determined to wean itself from Chinese imports, using both tariffs and incentives to build a domestic industry. In the US incentives are being provided to grow production. And always there are newer technologies that could supplant polysilicon as the main material, though they have struggled to gain more than 5 per cent share of the market thus far. The most promising of these, emerging from the laboratory in the last decade, is a class of materials known as perovskites.

Climate-change-related materials advantage is not limited to silicon and PV. A key technology strategy for climate change mitigation is electrification, particularly of transport. If the electricity generation is to be mainly from low/zero carbon sources, then replacing fossil fuels with electricity is an effective strategy. The materials challenges are mainly around storage in batteries and electrical components to make this all work.

The leading battery technology is lithium ion. Lithium is neither highly abundant (like silicon) or overly scarce. It is mainly found in various brines and salt flats, not surprising as it is chemically in the same family as sodium. Australia is the leading producer, with nearly half of world lithium production, and Chile has the most estimated reserves. There is little competitive advantage from lithium production per se, but making batteries is a more sophisticated process, and requires scale to achieve low costs. The possibility for competitive advantage in the future comes not from doing lithium-ion batteries better, or bigger, but from finding the advantaged battery technology that will replace them, at least for certain applications. While we use lithium-ion batteries today in cell phones, computers and electric vehicles, the future superior technology might be one that is best suited to one of these applications and not others.

Electrification also means more motors and more generators of electricity. A crucial constituent for modern motors and the generators of wind turbines are certain elements from the lanthanide series in the periodic table, sometimes known as rare earth elements, of which the highest demand is for neodymium, with cerium and lanthanum also being important. In fact, rare earth elements, with the exception of promethium, are not all that rare, as shown in this graph:[29]

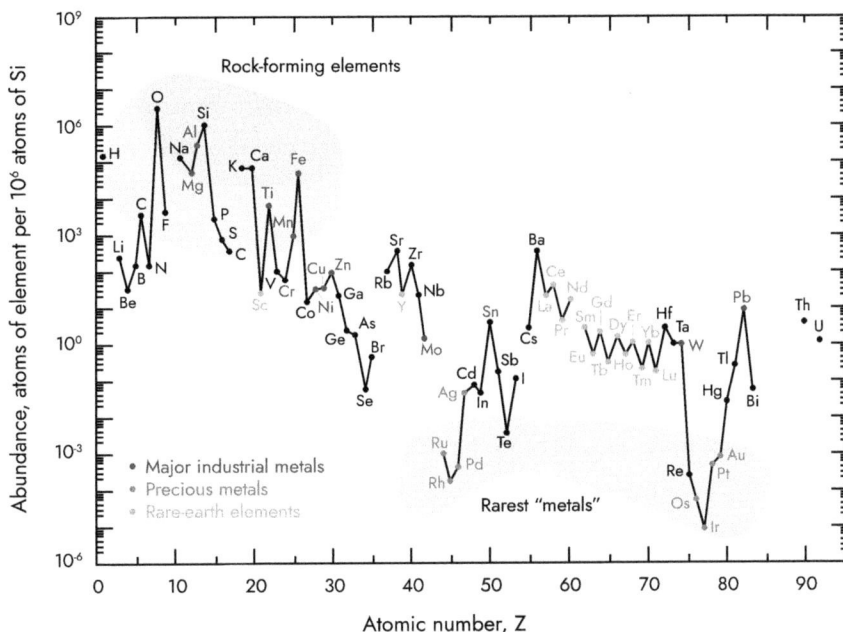

While there is relatively high abundance of these elements, the highest concentrations are in certain locations. At one time South Africa was the leading producer, then the US produced significant quantities, but in recent decades both have declined, so that in the 21st century only China and Australia are major producers. While this is a case of competitive advantage from possessing raw material, it may not be an enduring advantage. Rare earth elements are often found with coal deposits, and have, for decades, been discarded as part of waste. If economic processes can be found for recovering these elements from waste, the advantage that China has as a supplier may be eroded.[30]

The 'dream' technology for a low carbon future is nuclear fusion. It is easy to dismiss this as a possibility because it has been worked on for many decades without success, defined as getting more energy out than what is put in, and being able to do this for more than a few seconds. The most recent experiments in the UK and US have achieved something close to these 'breakeven' conditions, and only for very short periods. Fusion is a very difficult problem but one worth working on. A fusion reactor must tolerate a combination of very high temperatures and large quantities of radiation. As a recent article stated, 'The greatest problem faced in fusion isn't achieving the incredible temperatures required – it's the materials science required to maintain that environment long-term.'[31] Fusion is being developed through big international collaborations, such as ITER in Europe. It is unlikely that if the materials science problems are solved, they will lead to competitive advantage for any nation. However, if sometime in the coming decades it appears that fusion at commercial scale is possible, then it may replace a swathe of other technologies, leading to stranded assets and competitive disadvantage. This is not a problem that any business or nation is worrying about now.

Solar cell materials, lithium or other battery materials, and rare earths are but three examples of how the low carbon economy of the 21st century offers opportunities for competitive advantage to those countries with the science and engineering base to take advantage of them, along with the ability to provide the large sums of capital to develop the world-class enterprises that would have an impact.

## THE MATERIALS SHAPING OUR FUTURE

A lot of the solutions to problems of more productive use of major materials such as cement, steel and glass can be solved through better practices without any innovative technology. Durability and recyclability can often yield to design, again with minimal new technology required. However, if there is one lesson that we should take from the 20th century it is that technology always exceeds our expectations. It was a century that started with the automobile, the aeroplane, and electrification and ended with the mobile phone. One of the great American shop chains in the later 20th century was Radio Shack. As Steve Cichon observed after looking at a 1991 Radio Shack full page newspaper advert,[32] of the 15 electronic items offered there 13 are now done on your phone, as well as many more functions that were not in that advert. The only two missing are a detector for police radar and a large speaker to play music loud enough to infuriate your neighbours. The advert also tells you to look in the phone book for your nearest Radio Shack store.

How will technology for materials exceed our expectations in the coming decades, and how can this lead to competitive advantage for nations? The selection in the following sections is necessarily a personal choice, certainly not an exhaustive list of what is possible.

## ADDITIVE MANUFACTURING

The problem of waste from subtractive manufacturing was highlighted in the previous section. In additive manufacturing, a desired shape is built up rather than carved away. It is something like the difference between the work of a ceramist compared to that of a sculptor. The popular name for additive manufacturing is 3D printing. While it is already commercial, its impact is only beginning to be felt. If something can be made from plastic powder, steel powder, any sort of alloy, even concrete, it can be built up under computer control to a desired shape. Besides eliminating the waste from subtractive manufacturing, it facilitates adjusting the shape to meet specific requirements. Doing this for large numbers of manufactured objects is 'mass customisation', something that has been talked about for several decades and is now a reality.

One obvious target is joint replacements. Hip and knee replacements were one of the great surgical innovations of the 20th century, relieving millions of older people from decades of painful living or reduced mobility. Another major 20th-century achievement was three-dimensional imaging of any part of the body, both with X-rays (computer tomography or CT scans) and magnetic resonance imaging (MRI). Combine a CT scan of a hip joint with 3D printing and a joint replacement can be made in exactly the right size for the patient. A related, simpler application is for dental crowns and implants, where an optical scan may be sufficient to produce the specifications for a printable crown made in the dentist's office. The first machines to do this by subtractive manufacturing are already in use.

A less obvious application is machinery. Companies like Siemens and GE make large turbines in many different designs. These machines stay in service for many years, although new designs are introduced regularly.

From time to time a turbine blade breaks. Either the user has a stock of spare parts or must wait until the manufacturer can slot into the schedule to make a replacement. In practice a turbine may be out of service for six months from a broken blade. With 3D printing a stock of metal powder and the appropriate specifications can be kept regionally, so that when the need arises a new blade can be made on the same day. What an increase in productivity! The same principle could apply to parts for old Land Rovers, Harley Davidson motorcycles, clothes washers, and on and on.

In the early 20th century Thomas Edison thought he could mass produce houses from cast concrete. He built a few but the idea failed. Now there are companies such as SQ4D and Mighty Buildings controlling this process with 3D printing, making all the walls of a house in 48 hours of print time spread over eight days, with lots of design flexibility that was not possible with Edison, who was confined to the shape of moulds he had. Companies are also printing the light gauge steel for the house's interior structural support.

## Metamaterials[33]

Additive manufacturing is not, so far, about new materials but the much more efficient use of fairly mundane existing materials. There are radical new materials and more radical materials designs that are emerging and may come into wide use. One of these is a class known as metamaterials, the prefix meta (beyond) indicating that these are materials assembled in such a way as to have properties never found in the natural materials. The idea of metamaterials, building structures that have quite different interactions with electromagnetic waves, has been around for more than a century, dating back to the early experiments with antennas. The basic principle of what is being manipulated can be understood with a little elementary physics.

In normal materials, when light passes from one medium, for example air, into another such as glass, the light wave is bent because of the different velocity (v) at which light travels in the second medium. This is known as refraction, and is governed by what is known as Snell's Law as shown in the following figure:

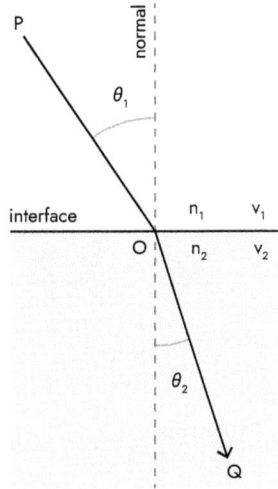

The two media are characterised for this purpose by what is known as the refractive index, n, and Snell's Law states that $\sin \vartheta_2 / \sin \vartheta_1 = n_1/n_2 = v_2/v_1$. Now, in a metamaterial, at least in some realisations of this, the behaviour of the wave compared to a conventional material is:

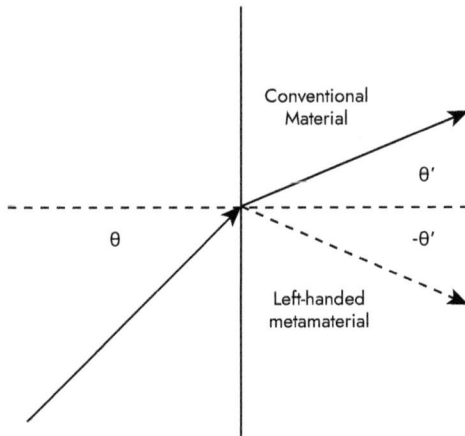

Examples of simple metamaterials.

For Snell's Law to be satisfied, the refractive index of the metamaterial would have to be negative.

The physics of how this is accomplished is quite complex, beyond the scope of this book. Structures consisting of several materials are constructed, a simple example being in this case made from copper and fibreglass.

The size of the individual cells in such an array must relate to the wavelengths of light with which it will interact. This makes it easy to see why the first metamaterials were designed to interact with radio waves or microwaves, with wavelengths of metres down to millimetres. Many of these have been made using 3D printing. For infrared or potentially for visible light, with wavelengths of about $10^{-7}$ metres, much more advanced manufacturing techniques are required.

There are numerous potential applications of metamaterials, but thus far only antennas are actually in use. Such antennas are more efficient and smaller than conventional designs. Other applications are for high absorption of light, cloaks of invisibility, metamaterials tuned to sound waves and similar frequencies both for sound absorption/manipulation and seismic protection, and for lenses having greater resolution than is even theoretically possible with conventional lens designs. Some of these are likely to be first seen (or not seen) in military applications, indeed some may already exist and be highly classified.

## Nano Nano Nano

*What I want to talk about is the problem of manipulating and controlling things on a small scale.*

Richard Feynman, December 1959[34]

On 29 December 1959 the great physicist Richard Feynman gave a lecture to the American Physical Society entitled 'Plenty of Room at the Bottom', in which he laid out the thesis that there was a lot to be done in both basic physics and materials science by working at much smaller sizes, i.e. manipulating materials at the nanometre (abbreviated nm, $10^{-9}$ or one billionth of a metre) level, say at the level of tens to hundreds of atoms. Moreover, he envisioned numerous applications that could come from such abilities. It took a little while for this to get going, but the last quarter of the 20th century and the first decades of the 21st have seen massive developments in nanomaterials and nanotechnology.

Suppose that instead of a semiconductor-fabricated device (see Tech Talk 5) for electronics the semiconductor was confined to a ridiculously small size, say 1–10 nm. It turns out that such confined semiconductor materials have quite different optical properties to larger scale semiconductors, as a result of effects understood through quantum mechanics. So unusual are these combinations of atoms at this small scale that they have properties as if they were different atoms entirely. Such materials, many of which have now been made, are called quantum dots.[35] They can function as media for transforming light into electricity (solar cells) or electricity into light (quantum LEDs). They are also finding application in medical imaging and in drug delivery. There are reports of novel photochemistry being carried out on the surface of quantum dots. Quantum dots are a large area of research and beginning to find commercial application; with unique optoelectronic properties they are likely to be in widespread use in coming decades. Moungi Bawendi, Louis Brus and Aleksey Yekimov shared the 2023 Nobel Prize in Chemistry for the invention of quantum dots.[36]

Nanoparticles open many other possible routes to new materials. By creating such particles from minerals, even mundane things like calcium carbonate, and then reacting with various biological polymers, it is possible to produce materials with properties of plastics but without any fossil fuel components.[37]

These few examples selected from a plethora of materials science nanotechnology directions provide an indication that some aspects of the materials and the techniques for fabricating them are going to play a major role in the materials that are important to our society in the 21st century. It may be that the most important ones are not yet out of the laboratory, however, the ecosystem of universities linked to venture capital and scale-up manufacturing is, in some countries, now sufficiently powerful that the time from laboratory to commercial application has been condensed.

## Carbon

Any first-year chemistry textbook of the 1960s would teach its readers that there are two forms of carbon – diamond, and graphite – with very different spatial and electronic structures, and hence different properties. Despite this seeming to be settled science, carbon remained an active area of research throughout the 20th century and continues to be so. In the 1960s there were predictions that carbon could form a stable ball structure containing 60 atoms, and this was discovered experimentally in 1985 by Harry Kroto, a visiting scientist from the UK working at Rice University, along with Richard Smalley and Robert Curl. It is likely that several other groups found $C_{60}$ at about the same time, but it was the Rice group that was most definitive about what they had made and a reproducible way for obtaining it. Similar carbon forms had also been deduced to exist based on light emitted from red stars, known as carbon stars. Because the structure of the $C_{60}$ ball consists of five- and six-membered carbon rings, like the pattern used by the architect Buckminster Fuller in his geodesic domes, the new form of carbon was given the name buckminsterfullerene, and it is now known colloquially as buckyballs.

Once it was clear that carbon would form stable structures other than diamond and graphite, many more were discovered. In 1991 the Japanese scientist Sumio Iijima published a paper describing small carbon tubes, now known as carbon nanotubes, and this started off a major area of research. It is now clear that many others had observed such nanotubes in the past, but for various reasons the work was not widely known. While buckyballs have not found any widespread commercial application, materials incorporating carbon nanotubes are already available in such things as tapes and sports equipment. The real potential for electronic devices, however, is still to come.

While not a new three-dimensional structure per se, the discovery by Andre Geim and Konstantin Novoselov in 2004 that it was quite easy to peel off a single layer of graphite, which they named graphene, and which was shown to have very interesting physical, chemical and electronic properties, was yet another new starting point for carbon materials. Over the next five years routes to producing graphene in quantity were developed. Graphene is already found in tennis racquets, touch screen displays, batteries, solar cells and in composite materials.

At about the same time as the graphene work, Xiaoyou Xu and co-workers at the University of South Carolina were looking at new methods for purifying carbon nanotubes and produced very small carbon particles ca. 10 nm in diameter that proved surprisingly stable. These are known as carbon dots, or sometimes as carbon quantum dots, because, like the semiconductor quantum dots described earlier, they have properties that can only be explained by quantum mechanics. These carbon dots have been shown to have low toxicity in humans, and can therefore find application in drug delivery, as well as in medical imaging, a result of their fluorescence properties when irradiated with certain wavelengths of light. Their surface can be modified in various ways, either to make it more passive (less reactive), or to bind molecules that give the carbon dots an active role, for example in enhancing photosynthesis in plants. Several start-up companies are trying to exploit the properties of carbon dots.

It was asserted not long ago[38] that by the end of the 20th century all the great problems of chemistry were solved, and there was not much new to discover, perhaps except for understanding heterogeneous catalysis. While it is true that predictions of novel carbon forms beyond diamond and graphite were made some time ago, and it is likely that carbon nanotubes were observed many decades ago as well, the continuing discoveries in such a chemically simple thing as pure solid carbon does make one question what more is out there.

## Steel, Concrete and Glass

More than half of all industrial emissions of $CO_2$ come from making the three essential structural building materials. Making one tonne of steel produces two tonnes of $CO_2$. Even wood, seen as a natural material, involves energy use because of the drying of timber used for construction. A lot of effort has gone into finding ways to use these materials more efficiently, effectively the dematerialisation strategy described earlier in this chapter. Optimising the amount of each in a building, that is, using concrete only where it is needed for compressive strength, not over-specifying the amount of steel or redesigning steel beams so as to achieve the same structural goal with less steel, these have been ongoing areas of progress. However, because the amount of energy required is so huge, if society is truly to achieve dramatically lower greenhouse gas emissions in the coming decades, radical innovations are required. While it has taken 30 years or more to lower the cost of renewable energy sources such as wind and solar, the changes to steel, concrete and glass need to be accomplished on much more rapid time scales.

The concrete challenge, from a historical perspective, looks extremely difficult. Even putting aside the fruitless 1,000-year search to duplicate Roman pozzolanic cement, it was another 300 years until Portland cement was invented, and that has dominated with only incremental improvements for nearly two centuries. Still, the developments in chemistry during the 20th century allow far greater understanding of solid-state reactions and structures. So-called geopolymeric cements,[39] which some have proposed to have been known to the Egyptians, have begun to find their way to market, and have much lower energy requirements than Portland cement. These are based around the structures of aluminosilicates, that is, minerals with aluminium, silicon and oxygen, which are quite common in clays and have special network properties that have found use in many other applications. A typical geopolymer structure is:[40]

$Al-O-Si$ ... $Si-O-Si$ ... $Si-O-Al$ $K^+$ ... $Al-O-Si$

[chemical network diagram of $Al$, $O$, $Si$, $K^+$ linkages]

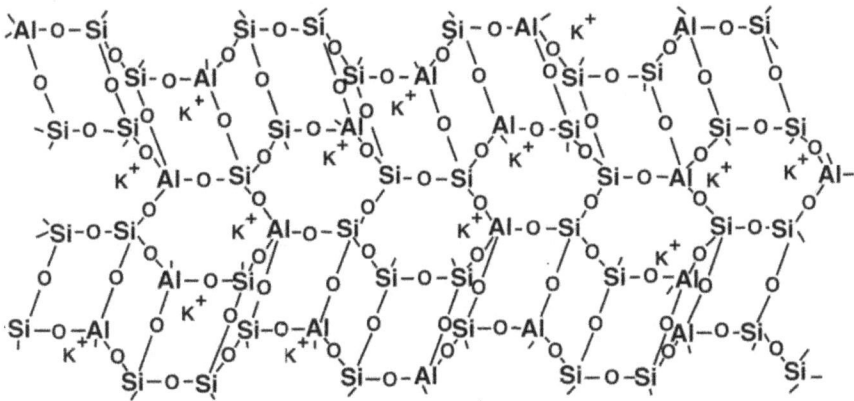

These are far from being zero carbon emissions when made into cementitious materials but are ca. 70 per cent better than Portland cement. This seems to provide a starting point for finding other materials.

Reducing the greenhouse gas emissions in steel processing may yield to novel approaches. As currently practiced, steel is a $CO_2$ emitter at almost every stage of the process, starting with energy used to pelletise iron ore so as to make it reactive enough for the iron oxide to be reduced to metallic iron in the blast furnace. Overall, the steel industry emits more than 3 billion metric tonnes of $CO_2$ every year, more than the weight of the steel produced. Most of the new methods being trialled simply replace fossil energy with renewables, where that is possible, use hydrogen for reduction, with the hydrogen coming from electrolysis powered by renewables, or use novel electrochemical routes from iron oxide to iron, again using renewable electricity.[41] At this time, all the pathways being considered would be at least 20 per cent more expensive than current steel production, though the main problems are that they are competing with others for off-peak electricity and hydrogen, and that they would require substantial capital expenditure. Still, as reference 41 points out, there may be innovations that cut the $CO_2$ emissions by 50–70 per cent; this still leaves a long way to go but would be a big contribution in the coming decade.

One of the approaches to reducing the embedded energy content from steel in buildings is to use more wood. The most popular material starting to find application is called 'Mass Timber',[42] and consists of multiple pieces of wood that are either laminated together or nailed at various angles to grain. These can substitute for steel beams and are strong enough to be used in 10-storey buildings. Mass Timber has been shown to retain its structural integrity even in fires, as it chars on the outside but does not burn through. It seems likely that architects, appreciating the appeal of wood's appearance as a building material, will begin to use Mass Timber. The key barrier to be removed is changing building codes, which in many regions do not permit wood in taller buildings.

While there appears to be little opportunity for reducing the embedded energy in glass, many things can be done to make glass a more environmentally attractive building material. Self-cleaning glass is already on the market, reducing the cost in both labour and materials for large structures with a lot of glass. Even more interesting is coating that makes the outside of buildings solar photovoltaic energy generators. Companies such as Onyx Solar[43] have specialised in this area, known as Building Integrated Photovoltaic, BIPV, and this is likely to become a routinely used building material in the coming decades. This 'glass' reduces the energy requirement for heating or cooling of the building, while generating some or all the remaining energy that is required.

## High Entropy Alloys

Alloys, one of the fundamental chemical tools in metallurgy, have been around for thousands of years. Somehow people learned that by adding (usually) small amounts of an element to a metal its properties could be altered in ways that were beneficial – strength, corrosion, friction, temperature stability, etc. The importance of alloys for steel and for various applications of aluminium have been described elsewhere in this book, as has the competitive advantage that can be derived from systematic investigation of alloys through a research programme such as that carried out by Alcoa on aluminium. Most alloys of commercial significance today are the result of the addition of between one and three elements to a pure molten metal.

It was common wisdom that mixing a larger number, say five or more, different elements to try to form an alloy would not work, because it was unlikely that a stable phase would result. When small amounts of a few elements are added, the alloying elements can fill vacancies in lattice, or perhaps trigger a switch to a slightly different crystal structure with more desirable properties. But with too many elements, you might just wind up with separated phases rather than a novel alloy material.

Work over the last 20+ years, particularly by Jien-Wei Yeh in Taiwan and Brian Cantor in Oxford, have shown that alloys of large numbers of elements are indeed possible. Because there are many different elements present, and they are not necessarily located in the same position in the crystal structure as one moves through the material, these have become known as high entropy alloys, i.e. they have a higher degree of disorder than more conventional alloys. It is this higher entropy which can, counterintuitively, lead to stability, as well as unique properties.[44] Yeh has developed explanations of the effects which lead to both stability and extraordinary properties of these new materials. While for millennia alloys were created by trial and error, this became more systematic with the development of chemistry and metallurgy, particularly in the 19th century. Now, with much better understanding of the fundamentals of atomic interactions, and with far more powerful computational tools, prediction of stability and properties becomes possible.

The impact of high entropy alloys might happen in some common material, like a new building material, but it is more likely to be found, at least initially, in applications which are currently struggling to solve the most extreme materials problems, such as nuclear fusion, where materials must withstand temperatures and radioactivity related impacts beyond anything now available. Since its inception it was clear that solving materials problems was crucial to making it work. Most of these problems remain after more than a decade of work. High entropy alloys, created by predictions from advanced computational tools, probably represent the most fruitful route to finding solutions, if they indeed exist.

## Superconducting Materials

A superconductor (see Tech Talk 5) is a material that undergoes a transition as temperature is lowered whereby its electrical resistance disappears. Such a material would allow current to flow over long distances without losses. Superconductivity was first observed in 1911 at extremely low temperatures, ca. -270°C. Subsequent research over many decades uncovered materials that were superconducting at progressively higher temperatures, up to ca. -235°C. Then in the 1980s there were a series of discoveries that led to superconducting materials at much higher temperatures, ca. -125°C. While the earlier materials required liquid helium to achieve the superconductivity, these newer ones could be superconducting at temperatures using the much more easily obtained liquid nitrogen or liquid air. These discoveries led to speculation that new materials that were superconducting at or near room temperature (even in a very cold room!) would soon be discovered. That has not been the case, except for a few materials at very high pressures.

If higher temperature superconductivity were achieved, and many scientists believe that this is possible, and these superconductors could be formed into long wires, this would revolutionise both the transmission of electricity and of information. For example, there are vast reserves of natural gas in Alaska and Siberia. These reserves could be converted to electricity using conventional power plants, the carbon dioxide produced could be reinjected into the gas reservoirs, and the electricity transmitted down to the US or the population centres of Asia respectively without loss. Likewise, information that is now transmitted over long distances by optical fibre might be advantageously transmitted via superconducting cables.

## Biology Meets Materials Science

Most of the plastic and synthetic fibres in today's materials world come from oil by-products. In a modern petroleum refinery, one that is integrated with chemical production, 80–90 per cent of the crude oil winds up in liquid fuels, some lesser amounts in bitumen/asphalt or other heavy products such as coke, and the remaining 10–15 per cent is for the petrochemicals business. With a move away from gasoline and diesel as transport fuels, refineries would have to be configured very differently to continue to produce the building blocks for packaging and fabric polymers. One way the industry has been planning for this transition is to make more of these polymers from natural gas rather than from oil as the starting material. Nonetheless, it is likely that the costs would be higher than they are today because the economies of scale of the giant refinery-petrochemical complexes would not be there.

This cost increase opens the possibility for other routes to be competitive. The huge advances in biotechnology in the last 50 years, and especially the last 20, have opened the possibility that biological routes could replace petrochemicals. Twenty years ago, large companies such as DuPont[45] and Cargill[46] were producing commercial materials from biological starting points, such as engineered *E. coli*. So far these have not taken large market shares away from traditional materials.

There are, however, large numbers of early-stage companies working at novel materials, and not all of them in the US, Europe, Japan and China either. For example, Bio-Eco[47] is a Thai company making considerable progress in bioplastic packaging using agricultural starting materials, such as sugar cane, corn and cassava. While taking corn and making it into plastic to package up corn for sale in a supermarket may not be very clever, there are huge quantities of waste biomass associated with most agricultural products. If processes can be developed to turn this into useful products that replace fossil-fuel-based materials, then there is something interesting.

Leather is also a target for novel materials. While it is already a biomaterial, leather production involves a lot of land, water and feed. The possibility of growing leather from mushrooms (mycelium) or producing it in a pure biological culture, for example using stem cells taken from an animal,[48] in much the same way as synthetic meat is being made, offers alternatives to traditional leather. There is a growing population that do not want to wear animal products and don't want to wear fossil-fuel-derived products. Approaches such as mushroom leather and lab grown vegan leather[49] may be a way out of that apparent dilemma.

These are only a few glimpses into biological routes to materials that are actively being pursued now. Aside from cost, they all suffer from the fact that biology, even at its most efficient, is extremely slow compared to traditional petrochemical routes. All of this is still at an early stage, and much of the scientific and engineering focus from biotechnology has been on pharmaceuticals, cosmetics and liquid fuels rather than on materials. As this changes, and pressure to replace single-use long-lived plastics with compostable packaging materials from non-fossil fuel sources grows, many advances can be expected.

## THE FUTURE AND COMPETITIVE ADVANTAGE

The examples presented in the preceding sections are but a few of the materials advances that could make for radical change in society. Many others have been speculated upon by various authors,[50] with varying degrees of documentation as to their likelihood. Some of these novel materials have already been realised in the laboratory, some are making first inroads in the market, and others have yet to be accomplished.

There are some common features.

- All are or will be the product of intensive, well-funded, frontier research programmes. Some of this research will be targeted to achieve a particular result such as a higher temperature superconductor, but in every case they will be underpinned by fundamental research into the underlying chemistry, physics and engineering of materials.

- Scaling from laboratory to pilot stage to industrial capacity will require the best engineering skills, and the faster one wants to achieve commercialisation, for example with a novel cement, the more skilled the engineers must be.

- A strong entrepreneurial culture must be present, one that encourages people not only to start enterprises but to build them to the scale where they impact society.

- There must be the ability to mobilise the risk capital required to bring a novel material and its manufacturing process to market for the first time. This may be state/government sponsored or use private capital, but it must have the appetite to take the risk and the willingness to make many bets, some of which will not pay off.

- Many of the changes in materials that are envisioned will be driven by environmental goals, such as reduced greenhouse gas emissions, elimination of plastic waste, and waste reduction in general. Societies embracing these goals, indeed seeing them as completely congruent with their own vision of the future, will be inclined to back every step of the transition from old ways of doing things to radically new ones.

At the very least these five points are required to achieve competitive advantage from materials enterprises. Of the 200 or so countries in the world today perhaps fewer than 20 have the potential to accomplish them. Certain materials or fields of application may be dominated by different countries, but it seems likely that at least some will achieve competitive advantage that is sustainable at least over a few decades, even if the historical mechanisms which sometimes resulted in centuries of competitive advantage are no longer possible.

18 https://www.lightmetalage.com/news/industry-news/automotive/aluminum-continues-unprecedented-growth-in-automotive-applications/ accessed 12 May 2021

## Notes

1 Michael E. Porter, *The Competitive Advantage of Nations*, Macmillan, 1990, Chapter 13

2 Howard T. Odum and Elisabeth C. Odum, *A Prosperous Way Down*, University Press of Colorado, 2001

3 Andrew McAfee, *More from Less*, Simon and Schuster, 2019

4 Julian M. Allwood and Jonathan M. Cullen, *Sustainable Materials, without the hot air*, UIT, 2015

5 Vaclav Smil, *Making the Modern World*, Wiley, 2014

6 Tim Jackson, *Material Concerns: Pollution, Profit and Quality of Life*, Routledge, London, 1996

7 Paul Krugman, 'Competitiveness: A Dangerous Obsession', *Foreign Affairs*, March/April, 1994

8 About the Council – Council on Competitiveness, www.compete.org accessed 4 May 2021

9 MPRA_paper_68151.pdf (www.uni-muenchen.de) 5 May 2021 https://mpra.ub.uni-muenchen.de/68151/1/MPRA_paper_68151.pdf

10 James Fallows, 'Japan', *The Atlantic*, 4 May 2021, https://www.theatlantic.com/author/james-fallows/japan/

11 See for example 'China Is a Paper Dragon', *The Atlantic*, 3 May 2021, https://www.theatlantic.com/ideas/archive/2021/05/china-paper-dragon/618778/

12 Given that optical fibre is strung in quantity from towers stretching across the landscape, this is again a steel and glass infrastructure.

13 List of countries by GDP (nominal) per capita – Wikipedia https://en.wikipedia.org/wiki/List_of_countries_by_GDP_(nominal)_per_capita accessed 11 May 2021

14 Vaclav Smil, *Making the Modern World*, pp119–155

15 www.label.averydennison.com accessed 13 May 2021

16 'How aluminium beverage can is made – material, production process, manufacture, making, used, composition, structure', https://www.madehow. com/Volume-2/Aluminum-Beverage-Can.html accessed 13 May 2021

17 US EPA, OLEM (2016–03–08). 'Sustainable Management of Construction and Demolition Materials', US EPA, https://www.epa.gov/smm/sustainable-management-construction-and-demolition-materials#:~:text=EPA%20 promotes%20a%20Sustainable%20Materials,mine%20and%20process%20 virgin%20materials accessed 13 May 2021

18 This is a very simple example of the law of unintended consequences. Make the landfill tax high enough, and some country will offer to take the waste for less than the tax bill, including shipping (which is a very inexpensive mode of transport).

19 Technology Strategy Board TSB – Designing Buildings Wiki, https://www. designingbuildings.co.uk/wiki/Technology_Strategy_Board_TSB accessed 13 May 2021

20 For an example of a company introducing such tools to the construction industry see https://qualisflow.com/ accessed 6 September 2021

21 An eloquent discussion of this from the consumer point of view is by the founder of Buy Me Once, buymeonce.com, Tara Button, *A Life Less Throwaway*, Ten Speed Press, 2018.

22 'Remarks at the University of Kansas', 18 March 1968, JFK Library, https:// www.jfklibrary.org/learn/about-jfk/the-kennedy-family/robert-f-kennedy/ robert-f-kennedy-speeches/remarks-at-the-university-of-kansas-march-18-1968 accessed 13 May 2021

23 https://ec.europa.eu/eurostat/statistics-explained/index.php?title=Waste_ statistics/es accessed 13 May 2021

24 Estimation of municipal solid waste generation and landfill area in Asian developing countries, https://www.researchgate.net/publication/50350008_ Estimation_of_municipal_solid_waste_generation_and_landfill_area_in_ Asian_developing_countries accessed 19 May 2021

25 See reference 6, and his many other books and articles, most notably *Prosperity*

*Without Growth: Foundations for the Economy of Tomorrow*, Routledge, 2017, and most recently, *Post Growth: Life After Capitalism*, Polity, London, 2021. For a complete selection of his books and articles see Tim Jackson :: Publications, https://timjackson.org.uk/about/publications/#books

26  For example, Jason Hickel, *Less is More: How Degrowth Will Save the World*, Windmill, 2021, Giorgos Kallis et al., *The Case for Degrowth*, Polity, 2021

27  Geoff Mann, 'Reversing the Freight Train', *London Review of Books*, pp27–30, 18 August 2022

28  A good review of the Chinese developments and US response is 'Solar's Supply Problem', *Chemical and Engineering News*, 19 September 2022, pp21–25.

29  Rare-earth element – Wikipedia, https://en.wikipedia.org/wiki/Rare-earth_element#Uses accessed 7 November 2022

30  As an example of work in this area, see DOE Awards $19 Million for Initiatives to Produce Rare Earth Elements and Critical Minerals | Department of Energy https://www.energy.gov/articles/doe-awards-19-million-initiatives-produce-rare-earth-elements-and-critical-minerals accessed 7 November 2022

31  'A Material Future for Fusion', *Chemistry World*, August 2022, pp48–52

32  'Everything From This 1991 Radio Shack Ad You Can Now Do With Your Phone', *HuffPost*, 19 May 2021 https://www.huffpost.com/entry/radio-shack-ad_b_4612973

33  The Wikipedia page for metamaterials contains many references that are useful for further reading. https://en.wikipedia.org/wiki/Metamaterial accessed 20 May 2021

34  For a pdf version of Feynman's talk, see R. Feynman, plentySpace.pdf https://web.pa.msu.edu/people/yang/RFeynman_plentySpace.pdf accessed 20 May 2021

35  'Quantum Dot – an overview' ScienceDirect Topics, https://www.sciencedirect.com/topics/materials-science/quantum-dot accessed 20 May 2021

36  Press release: The Nobel Prize in Chemistry 2023, https://www.nobelprize.org/prizes/chemistry/2023/press-release/ accessed 2 April 2024

37  S. Sun, L-B Mao, z Lei, S-H Yu, H. Coelfen, Hydrogels from Amorphous Calcium Carbonate and Polyacrylic Acid: Bio Inspired Materials for Mineral Plastics, *Angewandte Chemie*, 2016 https://doi.org/10.1002/anie.201602849 accessed 1 June 2021

38 B. J. Bulkin, *Solving Chemistry*, Whitefox, 2019

39 Tony Whitehead, 'Beyond Cement: Geopolymer Concrete Carbon-neutral alternative uses no Portland cement at all Geopolymer Concrete, A Carbon-Neutral Alternative to Cement', https://www.the-possible.com/geopolymer-concrete-carbon-neutral-alternative-to-cement/ accessed 1 June 2021

40 aluminosilicate – Google Search 1 June 2021

41 A good recent review of new approaches to steel manufacture, at various stages of development, is in 'The Race for Green Steel', *Chemical and Engineering News*, pp22–29, 2021 (CEN.ACS.org).

42 Experimental Tall Wood Buildings Material: Mass Timber, https://www.autodesk.com/design-make accessed 2 June 2021

43 Onyx Solar – Photovoltaic Glass for Buildings, https://onyxsolar.com accessed 2 June 2021.

44 A brief but useful introduction to high entropy alloys can be found in *Chemistry World*, March 2024, pp25–27.

45 For a description of DuPont's Sorona polymer: https://biomaterials.dupont.com accessed 21 June 2021

46 Cargill – Bioplastics News discusses some of Cargill's products and joint ventures, https://bioplasticsnews.com/tag/cargill/ accessed 21 June 2021

47 About Us, Bio-Eco, https://www.bio-eco.co.th/about-us?___SID=U accessed 21 June 2021

48 Lab-Grown Leather | VitroLabs Inc, https://www.vitrolabsinc.com

# DIGGING DEEPER INTO THE TECHNOLOGY BEHIND MATERIALS

The technology behind the transformation of raw materials into manufactured products is pretty amazing, and rather than send readers to a lot of different places to read about it, some of the key technologies are described in the following sections. This is a selection, probably what I think you might be interested in. Yes, there are chemical formulas in some sections, but it should be understandable by anyone with secondary school level science education.

## TECH TALK 1 THE MOST AMAZING MATERIAL OF THE SECOND HALF OF THE 20TH CENTURY?[1]

One of the most ubiquitous materials in food packaging today is a polymeric film, either clear or metallised with a coating of aluminium. In the developed world, any trip to the supermarket will mean that this material is encountered in almost every aisle. It is the clear plastic or white lid of yogurt, the metallised packaging of crisps and coffee, the thin plastic film on various ready meals and the top of packaged fruit and vegetables. It has excellent barrier properties, meaning that it keeps out gases that will lead to spoilage, while holding in volatile components in food associated with freshness. Sealing in these volatile components means that food odours are rarely encountered in the modern supermarket. It also has sufficient resistance to heat and microwaves that it can be left in place when ready meals are prepared.

This is also the material that is used in various sorts of thermal blankets, such as are often seen wrapped around people who have been exposed to cold or packed by mountain climbers with forethought. The metallised film stops heat loss from the body and keeps cold air out. It was used by NASA in various space missions to protect against radiation damage from cosmic rays.

Various other applications are in such diverse places as electronics, magazine packaging, specialised scientific equipment, confetti, nail polish glitter, agriculture and printing. The Wikipedia page on this material lists 65 distinct uses.

The material that is used in all these applications is the polymer that we usually know by its DuPont trade name, Mylar. The polymer itself is the most well-known polyester, polyethylene terephthalate (PET) (see Tech Talk 2), used as a fibre for clothing and as a blown resin for plastic bottles. In 1950 a process was developed at DuPont, and at about the same time in the UK at ICI, to convey particular properties to sheets or ribbons of this polymer. The process is strictly mechanical, not chemical, and it involves heating the polymer and stretching it to three to four times its length as it is pulled along the ribbon length. This orients the chains in one direction. Then, usually in a second step, the ribbon is stretched by a similar amount in the direction perpendicular to the first stretching, so increasing the width of the sheet. As a result, the chains are oriented along two directions. After the drawing along both axes the orientation is set by heating in an oven. The resulting polymer is thus called biaxially oriented polyethylene terephthalate, BoPET for short. Although the orientation process produces crystalline regions in the film, the size of the crystals is smaller than the wavelengths of visible light, so the film appears clear rather than the opaque appearance normally associated with polycrystalline materials. To prevent the sheet from sticking to itself as it is rolled up, a coating of microscopic mineral particles, such as silica, is applied to the sheets, giving a tiny bit of roughness to the surface.

Many of the long list of applications rely on a few properties of the Mylar films. Having been stretched and oriented, once cooled the film is very resistant to further stretching. The surface is also very easily coated with a great variety of metals. It is these two properties that make Mylar ideal for magnetic tape. After it is coated with magnetic material (initially some form of iron, more recently various cobalt alloys) data, sound and images can be recorded by magnetising the particles. This can be read back and transformed into the original. From a materials point of view, it is essential that the magnetic coating be stable on the surface, and that even with repeated playback the length of the tape does not vary, because any lengthening or shrinkage would severely distort the recorded information.

The ability to metallise is also important for many of the food packaging applications. PET is a very good barrier polymer on its own, hence its use for many beverages, keeping oxygen out and carbon dioxide in. These barrier properties are further enhanced by the metallic film.

Many of the applications are more frivolous than those just discussed. Metallised BoPET balloons, wrapping paper, confetti and nail glitter do not advance any of the goals of sustainable development. While BoPET itself, as polyester sheet, is recyclable, metallised BoPET, which dominates applications, is not – being neither a pure polymer nor a pure metal product. It is possible that future recycling technologies may be able to overcome this limitation in a cost-effective way, but today most metallised BoPET winds up in landfill or is thrown away on the side of the road and ends up in the ocean.

## TECH TALK 2 POLYMERIC MATERIALS IN CLOTHING

*Synthetic Fiber*

> *Wallace Hume Carothers assignor to EI DuPont de Nemours & Company*
> *This invention relates to new compositions of matter, and more particularly to synthetic linear condensation polyamides and to filaments, fibers, yarns, fabrics and the like prepared therefrom.*
> Patent 2,130,948 for what became known as Nylon

The fibres in clothing are all long-chain molecules (polymers). Until the 1930s these were all biological (natural) polymers, but with the invention of synthetic polymers, such as Nylon, Terylene, Dacron and Orlon, clothing today is a mixture of natural and synthetic materials. The composition of synthetic materials can be tightly controlled in an industrial process, whereas for materials from plants or animals there will be significant variability and seasonality. Plant and animal fibres also need to be purified in a set of steps to separate them from other components, or to carefully retain minor components that give them special properties while discarding others.

## NATURAL MATERIALS
### Linen from Flax

Fundamental to many of the plants used in clothing (and in other materials such as paper) is the long chain natural polymer cellulose. It consists of the sugar component glucose formed into long chains, bonded through the oxygens at either end of a hexagon of carbons, and these chains are aligned with one another by what are known as hydrogen bonds, shown in the figure below as dotted bonds between chains.

For the fibres from flax from which linen is made, in addition to cellulose there is hemicellulose, a more complex polymer containing some of the same glucose chains, but also other sugars and branches, so it does not have the linearity of cellulose. Flax also contains lignin, which is either a relatively small molecule or can grow into large polymers, and is an important component of wood, as well as pectin, a related polysaccharide (sugar polymer), which becomes important in many food preparations as a thickener.

It is this complex, naturally occurring mixture of cellulose, hemicellulose, lignin and pectin (as well as some other minor components) which give linen its characteristic properties of texture and ability to be dyed.

### Silk

In contrast to linen, where the main components are polymers of sugars (polysaccharides), silk is made up of polymers of amino acids,  such as glycine, serine, and alanine, which are proteins. The main silk polymer is called fibroin. Once again hydrogen bonding allows the fibres to be aligned with one another, in this case forming sheets. Outside the fibroin is usually found another polymer in silk, sericin, which has much more serine and is a sticky coating. The silk polymers have extraordinary strength.

## Wool

Wool also consists of an amino acid polymer, keratin. While there is still hydrogen bonding between chains, the key feature of keratin is that some of the amino acids contain sulfur, and two chains get linked together by S-S bonds, with each chain curling up in a helical structure. In contrast to silk, wool is a relatively weak polymer.

In addition to keratin, wool contains significant amounts of a waxy/oily material called lanolin, which is itself a mixture of several chemical species. Lanolin is generally removed from the wool now and finds uses in pharmaceuticals and various skin treatment products, but in older wool fabrics it was used to achieve water repellent properties that were desirable for outerwear and especially for woollen sails.

## Cotton

The chemical composition of cotton is pretty simple – it is usually >90 per cent, often as much as 95 per cent, cellulose. There are small amounts of pectin, wax and protein consistent with the plant origin of the fibres, but all the key properties – strength, feel, washability, dying – are determined by the cellulose.

## Rayon

Rayon was developed in France in the 1860s as an alternative to silk, and it is just purified cellulose that has been spun into fibres. The cellulose mostly comes from wood, but some of it is made from waste in cotton production.

## Leather

While most clothing requires fibres that can be spun and woven, leather possesses different material properties. Nonetheless it is chemically related to other natural polymers discussed above, in that it also is a protein-like material. The amino acid sequence of a typical collagen is chains forming into triple helix structures, and these then aligning, a superstructure which leads to the properties of leather.

## SYNTHETICS

Some synthetic textiles and leathers were probably made during the 19th century, primarily by accident, as were other plastics. Chemists, trying to synthesise some small molecule, found that their reaction vessel was filled with a solid mass of material. But the great advances in synthetic fibres that revolutionised the clothing industry in the 20th century came because of fundamental understanding of the chemical structures of the natural polymers, and systematic reaction chemistry to replicate these at higher purity.[2]

### Nylon

In trying to develop the chemistry of polymers for practical applications, one of the first goals stated by industrial scientists was to make synthetic silk.

Wallace Carothers was a brilliant young chemist, recruited from Harvard by DuPont to do fundamental research on polymer science. He began by extending the work done by Hermann Staudinger in Germany, and with modest goals to make synthetic polymers of molecular weight of about 10,000. Soon, however, his bosses at DuPont encouraged him to try to find some new materials that would impact the business. In the 1930s Carothers and his team made the first truly synthetic textile fibre, named Nylon. This built on knowledge of the amino acid polymers that make up silk, but as a synthetic the chemical composition is controlled in the laboratory rather than in a silkworm. The bonds between the repeating units are 'amide bonds', which are the same as those between the amino acids in proteins. Part of the motivation for Nylon was to stop reliance on Japan for silk.

### Acrylic Fibres

Not long after the introduction of Nylon, DuPont launched a completely different type of synthetic textile fibre made by polymerising acrylonitrile (vinylcyanide), and this is called Orlon. This does not build on any biological models, and is a pure synthetic chemical fibre, designed by chemists to achieve certain properties.

## Polyester

Even before his synthesis of Nylon, Carothers was experimenting with making polyesters. In these syntheses instead of amines such as used in Nylon, he used glycols $HO-(CH_2)_n-OH$ to bind to the acid. When the team made Nylon they set aside the work on polyesters, because the Nylon seemed to be a much more promising fibre. Later, work in England at the Calico Printers Association in Manchester synthesised polyethylene terephthalate (using the glycol above where n=2) and having the following structure:

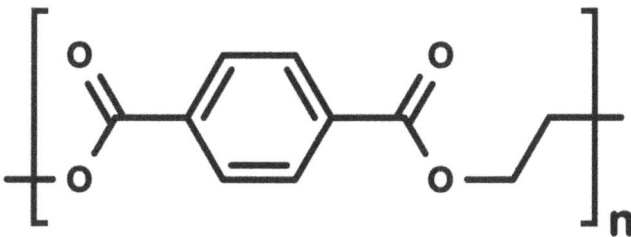

This polymer had the trade name in England of terylene, and DuPont developed it as Dacron. While Nylon became a dominant material for stockings (and many other non-clothing applications) Dacron and other polyesters found their way into all sorts of clothing, especially perma-nent-press suits.

## Lycra

Rubber also had a place in garments, particularly various undergarments, but natural rubber is not a great material for this purpose. It does not react well to perspiration, it is difficult to launder without degrading its proper-ties, and it is uncomfortable to wear. Most of the effort on synthetic rubber was motivated by needs of transport, but one spectacular development occurred in the late 1950s at DuPont, with the synthesis and commercial-isation of what became known as Spandex or Lycra.

It is made by the following reaction and can be seen to be a part of the family of polymers DuPont made and commercialised around amide linkages. Lycra can be stretched to double its length and return, repeatedly, to its original shape.

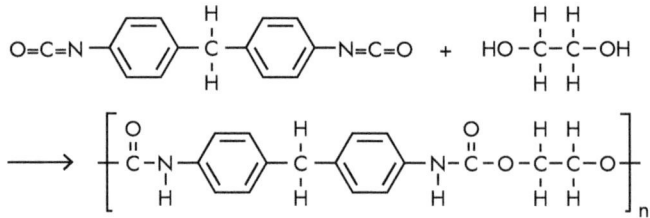

## Nomex and Kevlar

In the 1960s DuPont groups led by Wilfred Sweeney and Stephanie Kwolek took the Nylon structure of a polymer with amide linkages two steps further, using aromatic rings in the central amine structure rather than the aliphatic in Nylon. In this way they first created a polymer marketed as Nomex in which the rings are linked two carbons apart, which chemists refer to as a *meta* linkage, and then in 1965 the version where the rings are linked at opposite ends, a *para* linkage, which is marketed as Kevlar.

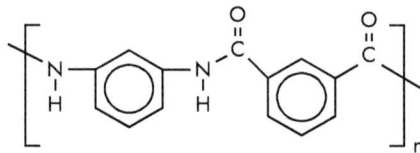

These synthetic polymers have remarkable properties, Nomex for its flame retardancy, leading to its use in all professions from firefighting to race car driving where protection from fire is required, and Kevlar for its extreme strength, in clothing applications mainly used for bulletproof vests.

## TECH TALK 3 THE INDUSTRIAL REVOLUTION IN WEAVING

Weaving is in essence a very simple idea. Once long continuous threads or fibres have been made, and manipulated into some sort of spools for handling, a series of threads are held in tension in the longitudinal direction (warp, the length of the cloth) while a second thread is passed over and under, back and forth (the weft) to produce a fabric of a given width.[3]

Carrying this out manually requires no complex technical knowledge, but it is tedious, back-breaking work. Nonetheless, that is how it was done for millennia.

As sources of power from machines became available in the 18th century, and as the British businessmen saw the need to eliminate bottlenecks holding back growth of textile manufacture, weaving became an obvious target for some form of automation. There are two obvious things required. The machine must hold alternate warp threads apart up and down so that the weave can be carried out, and the weft thread must be thrust through in one direction, then back again, repeatedly, without a human hand being required.

The weft yarn is wound up on a device called a shuttle; the challenge is to pass this thread from the shuttle through the warp and back again. For the weavers in Lancashire, England, in the early 1700s, two people worked each loom. One passed the shuttle through, the other caught it and passed it back.

John Kay was born in 1704 into a Lancashire family that was in the wool business. When he was very young his father died, and as the fifth child he had little inheritance. At home and as an apprentice he learned how looms worked, but his inclination was as an inventor, engineer and ultimately businessman. He began by making small improvements in the weaving process, but around 1730 he tackled the problem of the shuttle. His essential invention was to put the shuttle on wheels, with mechanical connection to the loom operator. By a mechanical action the weaver could trigger a lever to send the shuttle across and subsequently a second lever to send it back. All with proper alignment. He also improved the boards that kept the threads separated, and all the mechanics to ensure a tight and even weave. The key invention, however, was the shuttle, which Kay called the wheeled shuttle, but which became known as the flying shuttle.

John Kay's flying shuttle.[4]

The flying shuttle immediately resulted in a halving of the labour requirement for weaving, or, where one person was controlling the shuttle at both ends, it allowed for cloth wider than the reach of a human operator to be produced. Kay's invention was not a power loom, that would come later, but it incorporated the technology which could then be transformed through use of engines coupled to the loom. Indeed, the flying shuttle loom remained the dominant design for centuries after.

By speeding up weaving, Kay effectively moved the bottleneck from the loom back to yarn spinning. It took several decades more for this to be solved, through the invention of the so-called spinning jenny by James Hargreaves in 1764, also in Lancashire. If an engineer looks at a device such as a spinning wheel, they see all the incremental improvements that can be made to improve quality, reliability, speed and safety. The spinning wheel had evolved over centuries in this way to produce high-quality yarn from fibres. When an inventor such as Hargreaves looks at this evolved spinning wheel, it is in a completely different way. Seeing the spool rotating around a horizontal axis as the thread is wound up, he turns it vertically. Now when it is horizontal it was not possible to have multiple spools, because the threads would inevitably interfere with one another. But as soon as it was vertical, Hargreaves saw that he could increase from one spool to eight. Once the spinning wheel is set up, assuming an adequate supply of fibre, a single operator can produce eight times as much thread.

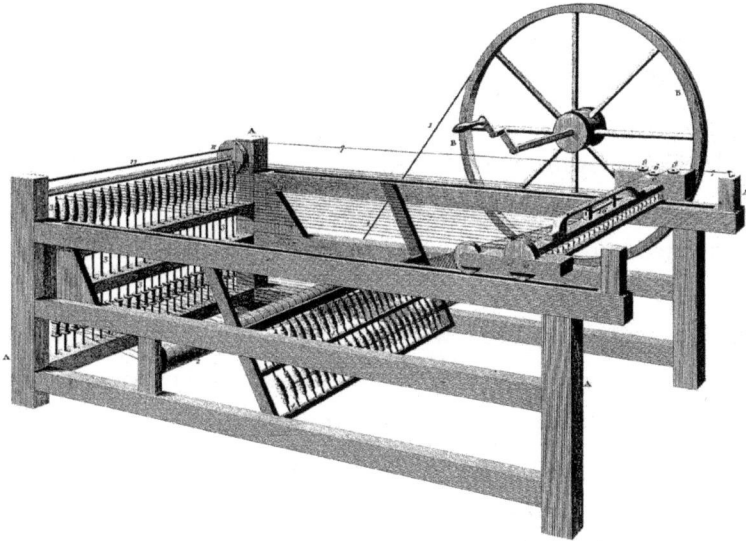

Spinning jenny.[5]

Eight spools were only the beginning. Over time Hargreaves produced machines where a single operator could control 120 spools.

Both Kay and Hargreaves were subject to frequent, violent attacks on their workshops and homes, by weavers whose livelihood felt threatened. Step changes in productivity, as discussed in Chapter 2, are never easy.

Engineering combined spinning with waterpower, when in 1769 Richard Arkwright produced the water frame, a water-powered spinning wheel. It again reduced labour requirement, but also provided more power to the wheel than a human operator could, so produced better thread. Arkwright's device did not apply waterpower to the spinning Jenny, so only worked on one spool. This was done in 1779 when Samuel Compton combined both devices, in what was known as the spinning mule.

Inevitably (from a systems-thinking perspective), just as weaving improvements stimulated spinning inventions, greater spinning productivity meant there was a demand for more efficiently produced fibre. Cotton fibres are formed by the plant in the seed pods, the cotton bolls. The fibres are mixed with the seeds, and separating them was a slow labour-intensive process, occupying a substantial portion of the slaves on southern US plantations. A single worker could separate one pound of cotton from seeds in a day.

The archaeological record from India shows that there were various attempts to build devices (called cotton gins, gin being short for engine) for separating fibres from seeds. Most of these involved some sort of roller and stone, but they required considerable operator skill so that the seeds were separated but not crushed, and as a result they did not achieve widespread use.

Eli Whitney, born in 1765, was from a New England family. His mother died when he was young, and his stepmother did not believe in getting him educated. From a young age he showed aptitude in mechanical things, and by the age of 14 was running a nail factory on his father's estate. He worked to earn money, first so that he could get a secondary education, and then to attend Yale University from which he graduated with high honours. After graduation he could not secure a position, so decided to travel to the southern US, thinking to either teach or find some employment where he could make money. He went by ship along the coast, on board meeting a woman, Catherine Greene, who had a plantation in Georgia, and who invited him to come to Georgia with her. Her plantation manager was also a Yale graduate. Within weeks of his arrival Whitney saw the problem of separating cotton fibre from seeds and began to build a model of a device to do this, completely different from anything previously used. The basic idea was to push the fibres through a wire mesh, through which the seeds could not fit, and grab the fibres with metal hooks to pull them through. To make this into a 'process' the metal hooks were on a roller, so that as the roller was turned, they grabbed fibres from the mesh and deposited them on the other side. The cotton gin increased the productivity of a single labourer from 1 pound of cotton a day to 55 pounds! Whitney patented his invention, but it was too easy to copy, and he was not able to build a successful business around it. His business model, much in use today in other areas, was not to sell the gins but rather a service for farmers. Although his name is always associated with the cotton gin, he went on to other manufacturing businesses, particularly various sorts of guns, in which he was quite successful. This manufacturing business, carried out back in his native New England, increased the prosperity of the north.

It has sometimes been supposed that all this automation contributed to the end of slavery. Quite the contrary. The demand for cotton increased very rapidly and drove forward the economy both in Britain and in the South. As a result, there was a consequent increase in the need for slaves to farm larger cotton plantations.

## TECH TALK 4 PAPER FROM WOOD

*'You can make anything from lignin, except money'*
Paper industry folk wisdom

Cellulose and the related polymer hemicellulose are the most abundant biological substances on the planet, closely followed by the polymer lignin which gives trees their structural strength, as well as protecting the cells from external degradation. Lignin is essential to the tree, but undesirable if the cellulosic fibres are to be made into paper products. The essence of all papermaking processes is to separate the lignin from the cellulose, leaving the fibres intact. For such process to be economic, the chemicals used to break down the lignin must be recovered and at the very least the energy contained in the lignin must be used to provide heat for the process. Ideally something more valuable than heat could be derived from the lignin. Trees also contain varying amounts of fats and resins, depending on the type of tree. Ideally these can also produce a by-product of value.

While the chemical structure of cellulose is quite simple that of lignin is very complex, and there are many 'lignins', depending on the tree (or even in grass). Despite much effort, the chemical structures of lignin have a lot of variability which has not yet been completely characterised, though there is some understanding of how the organic chemistry differs between, for example, hardwoods and softwoods.

The earliest methods used to separate lignin from cellulose involved mechanical hammering or milling of the chipped wood until the watery pulp separated, hence the expression 'beaten to a pulp'. Today, almost all of this is done chemically, mostly by what is known as the Kraft process. In the Kraft process, developed and commercialised in the late 1870s, the wood (may include bark) is cooked in a strongly alkaline mixture of sodium hydroxide (NaOH) and sodium sulfide ($Na_2S$) at 170° for about two hours. When these two are mixed the aqueous phase contains OH⁻ and SH⁻ ions, and these are highly active in cleaving bonds in the lignin structure, so that the polymer is broken down into small, low molecular weight fragments. They also react to leave the organic fragments with a lot of OH groups attached. This makes these fragments much more soluble in water than they would normally be, so they can be separated from the cellulose. The resins and oils float to the top of the aqueous phase, where they can be skimmed off, producing a product called tall oil. The lignin content decreases rapidly for the first 90 minutes, then its decline begins to slow. The process is optimised so that the lignin is reduced from ca. 25 per cent to <4 per cent, but not lower as longer cooking times start to degrade the cellulose.[6]

The chemical reactions involved tend to leave the pulp with a brown colour. This is the characteristic colour associated with packaging material. To make white paper, the pulp is bleached with some combination of usual agents such as chlorine, oxygen, ozone, sodium hypochlorite and hydrogen peroxide.

The liquid phase after the cellulose is removed is called black liquor. This contains the reacted pulping chemicals as well as the organic matter. Pulp mills have a process stage known as the recovery boiler, where the chemicals are regenerated for use in the process, at quite high efficiency, and the organic matter is burned to produce the heat necessary for running the process.[7]

## TECH TALK 5 CONDUCTORS, SEMICONDUCTORS AND INSULATORS

Well before the great industrialisation of electricity production and use, the electrical properties of materials were studied and classified, though certainly not understood at the level of atomic theory. That understanding was a 20th-century achievement. Demonstration of the flow of electricity through wires and some other natural fibres was done by Stephen Gray in England in the 1720s, who was also able to show that other materials were insulators – electricity could not flow through them. Gray was also the first to demonstrate the phenomenon of electrical induction, that is, that charge could flow from one body to another through air. Gray's associate John Theophilus Desaguliers, who was an experimental assistant to Isaac Newton, was the first to use the names conductor and insulator. The inverse measure to conduction in a material is resistance, so insulators such as Bakelite (Tech Talk 11) have very high resistance, and conductors such as copper very low. The difference can be as much as 24 orders of magnitude, that is, 10 to the power 24. By comparison, the difference between a good conductor like copper and a somewhat less good one, for example aluminium, are relatively small:

| | |
|---|---|
| Silver | 62,1 |
| Copper | 58,7 |
| Gold | 44,2 |
| Aluminium | 36,9 |

Without a course in quantum mechanics, it is almost impossible to understand these electrical properties. But… atomic nuclei are surrounded by electrons, and except for the noble gases (helium, neon, etc.) these have varying inclinations to bind to other atoms, hence chemical bonds. Once the electrons are in these bonds, they are quite localised to the region between the nuclei. Since electrical conductivity is the flow of electrons most such materials are insulators. We describe the electrons as being in discrete states, energy levels, and in these states each electron has associated with it a set of quantum numbers that describe it. The Pauli Exclusion

Principle states that no two electrons can have the same set of quantum numbers, so once the possible combinations are used up any additional electrons must be in a different energy level.

In a metal, the energy levels fall very close together, so close that rather than appearing as a discrete state they appear and behave like a band of levels. If the number of electrons available to populate the band is not sufficient to fill it, the electrons can have increased mobility – effectively they can move between levels and are not spatially localised as in a simple covalent chemical bond where generally all positions are taken. When a potential difference is applied across such a material, so that there is a negative and a positive end, the negatively charged electrons will flow from one end to the other. In fact, it is best not to think of this as a flow of material but as the transfer of energy between electrons across the material, because the band represents energy levels.

With most metallic conductors, as the temperature is raised the resistance (designated by the Greek letter rho, $\rho$) increases while the conductivity (usually designated by the Greek letter sigma, $\sigma$, decreases, as shown in this figure:[8]

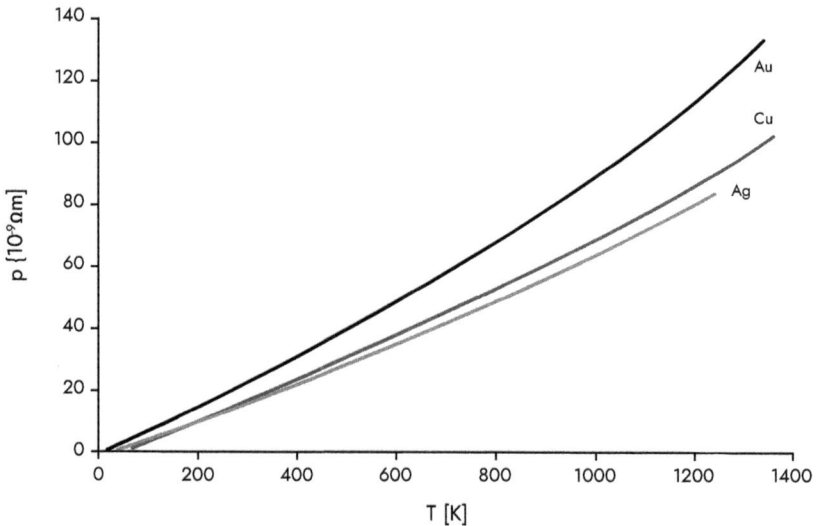

Here temperature is given in degrees Kelvin, the absolute scale of temperature ($0°C = 273$ K).

The generality of this temperature dependence of electrical conductivity in metals was crucial to the discovery of another class of electrical conductors, known as semiconductors. Their conductivity is less than that of metals, but greater than that of insulators. That they were in some fundamental way different was first observed by Michael Faraday in 1837, who found that modest conductivity of a material, silver sulphide, increased with increasing temperature rather than decreased. Some decades later this same effect was observed for selenium, and it was that observation which opened the way for the invention of devices making use of the properties of semiconductors.

If conductors have a partially filled band of energy levels which allow for electron flow, and insulators have discrete energy levels that are filled with tightly bound electrons, then, in this same model, semiconductors have a filled band, with another band of slightly higher energy into which some electrons can be promoted by heat or light. The energy difference between these bands is called the band gap. The conductivity of metals decreases with increasing temperature because of movement in the lattice disturbing the flow of charge, but for semiconductors the increasing temperature promotes more electrons into the conduction band, hence increasing conductivity.

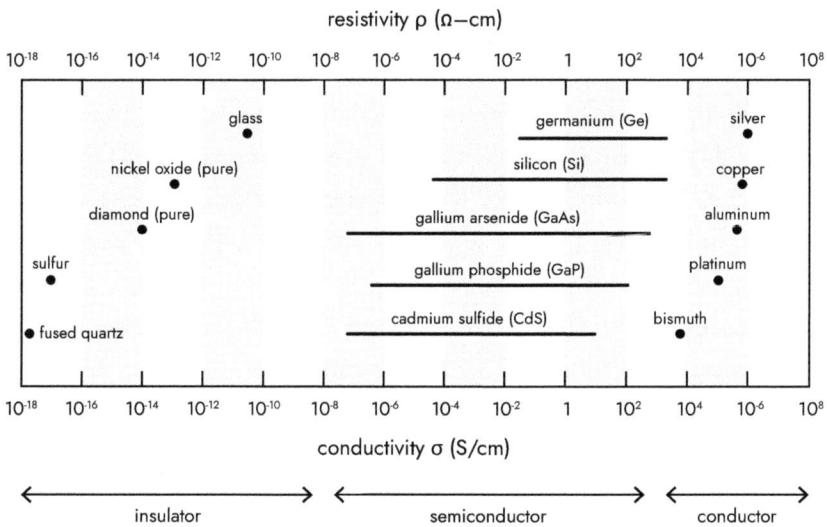

resistivity $\rho$ ($\Omega$–cm)

| $10^{-18}$ | $10^{-16}$ | $10^{-14}$ | $10^{-12}$ | $10^{-10}$ | $10^{-8}$ | $10^{-6}$ | $10^{-4}$ | $10^{-2}$ | 1 | $10^2$ | $10^4$ | $10^{-6}$ | $10^8$ |

glass

germanium (Ge)

silver

nickel oxide (pure)

silicon (Si)

copper

diamond (pure)

gallium arsenide (GaAs)

aluminum

sulfur

gallium phosphide (GaP)

platinum

fused quartz

cadmium sulfide (CdS)

bismuth

| $10^{-18}$ | $10^{-16}$ | $10^{-14}$ | $10^{-12}$ | $10^{-10}$ | $10^{-8}$ | $10^{-6}$ | $10^{-4}$ | $10^{-2}$ | 1 | $10^2$ | $10^4$ | $10^{-6}$ | $10^8$ |

conductivity $\sigma$ (S/cm)

insulator     semiconductor     conductor

Of the semiconductors shown in this figure, one of the best, and by far the cheapest, is silicon. Its wide application during the 20th century to devices suitable for computation revolutionised modern society, so much so that the pseudo-Latin term, 'in silico', for doing things via computer rather than physically, came into widespread use.

Pure elements and some chemical compounds can exist in different crystalline forms, and the electrical properties may vary. The most common example is carbon. As diamond it is an insulator, but as graphite it is a semiconductor.[9] Another interesting example is hydrogen. In its stable form we find it as the molecule $H_2$, a classic chemical bond, so that even in its liquid form at very low temperatures it is an insulator. But under very high pressures hydrogen can become a metallic substance and a conductor. It is believed that the hydrogen in giant gaseous planets such as Jupiter is in this metallic form.

While some materials are natural semiconductors, others that are insulators when pure, such as crystalline silicon, can be made into semiconductors by adding a small amount of another element with the correct electronic structure. This is called doping, and there are two types.

In N-type doping, phosphorous or arsenic are added. These elements introduce excess outer electrons that can't bind to the silicon lattice, and hence induce conductivity in the lattice. It is called N doping because of the negative charge of the electrons, that is, one is adding excess negative charge to the silicon. In P-type doping, elements such as boron or gallium are added. These have fewer electrons than are needed to bind to the silicon lattice, and create what are called holes, a centre of positive (P) charge. As it turns out, holes also induce conductivity. One can think of a hole as attracting an electron from a neighbour, then this moves the hole over to where the electron was, in effect leading to a flow of electrons. The effect is quite dramatic. Adding about 10 atoms of boron per million atoms of silicon increases the conductivity a thousand-fold.

While neither N nor P type semiconductors are particularly interesting in themselves, they become of great importance when combined in devices. Putting an N and P material together (a PN junction) in an electric circuit forms what is known as a diode. It blocks the flow of current in one direction but allows it in the opposite direction. When three such materials are put together, as say PNP or NPN, it is the basis of a transistor.

Properly configured in an electric circuit, transistors can act as switches or amplifiers. It is this which is the basis of all modern computers and other electronic devices. The ability to manufacture these in high density, connected into complex circuitry, is one of the greatest triumphs of engineering in human history.

## TECH TALK 6 HIGH-PURITY SILICON

One possible route to manufacturing advantage from a material is when the critical factor determining properties is purity. This can be the case with steel, for example. It is not that difficult to make molten iron from various ores, but if a manufacturer can produce iron with only trace amounts of carbon and other metals, then there is full flexibility to produce any sort of steel by adding back the desired alloying metals and concentration of carbon.

The extreme case of purity as a necessary attribute is silicon for electronic and photovoltaic applications. Silicon is very abundant in the earth's crust, second only to oxygen. It is found almost entirely as silicon dioxide, $SiO_2$, the main constituent of sand, quartz and other mineral forms. There are also other silicon-containing minerals, including those where it is found together with aluminium.

Silicon as a pure chemical element was first isolated by the great Swedish chemist J. J. Berzelius in 1824. It has a melting point of 1410°C and a boiling point of 2355°C. To make it, the quartz mineral is heated at 2000°C with coke (coal that has been roasted in the absence of air) so that the reaction $SiO_2 + 2\,C \rightarrow Si + 2CO$ takes place. In this high temperature process, some of the elemental silicon reacts with the carbon to produce silicon carbide, SiC, which is itself a useful product, the abrasive material in sandpaper. To maximise the yield of silicon, reactor conditions are adjusted so that the SiC reacts with additional $SiO_2$

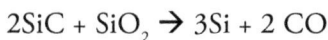

$$2SiC + SiO_2 \rightarrow 3Si + 2\,CO$$

The silicon produced in this way has a purity of ca. 99 per cent and is known as metallurgical silicon. It is used as an alloying material in certain steels and aluminium products.

For electronic applications, this level of purity is not nearly enough. The 1 per cent impurity, which can include elements like germanium and boron, has a major effect on the electronic properties. Silicon as an insulator is the base of most computer chips, and it is crucial that electronic isolation can be maintained between the components on the chip.

To achieve higher levels of purity, the solid silicon is converted to a gas by reacting with hydrochloric acid

$$Si \text{ (99 per cent)} + HCl \text{ (gas)} \rightarrow SiHCl_3 \text{ (gas)}$$

As a gas this is easily purified by distillation, so that impurities like boron can be separated. The gas is then decomposed at ca. 1150°C via the reaction

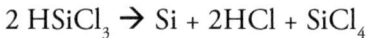

$$2\ HSiCl_3 \rightarrow Si + 2HCl + SiCl_4$$

The $SiCl_4$ reacts further with the HCl to form more $HsiCl_3$.

This purification method, including the method for depositing the pure silicon on a silicon seed crystal, is known as the Siemens process. It yields silicon of 99.99999 per cent purity, known as 7N (seven nines). This is the material used for most silicon photovoltaic (solar) cells.

For electronic applications, even higher purity is required. Further purification is done by modifying the distillation steps in the Siemens process. In this way 9N-11N silicon can be produced. The Siemens process is energy intensive and is best carried out in places with access to low-cost electricity. Several alternative approaches have been tried over the years, using lower energy, but each time improvements were made in the Siemens process that eliminated the advantage of the challengers. The biggest plants producing high-purity silicon are in China.

A final step is used for certain applications. The silicon produced by the Siemens process is polycrystalline, that is it is pure, but it contains many small crystals. The electronic properties are affected by the existence of this degree of disorder. To eliminate these 'grain boundaries' the silicon is melted, and a perfect crystal is lowered into the bath. This is slowly withdrawn, a technique known as the Czochralski Method after its Polish inventor, cooling slowly to form a monocrystalline material.

## TECH TALK 7 THE CHEMISTRY OF CEMENT
BY VICTORIA STOMBERG

Limestone is a widely occurring mineral, which is primarily calcium carbonate (in mineral crystal forms known as calcite and aragonite). It also contains variable amounts of magnesium carbonate, usually on the order of 3–6 per cent, which together with the calcium carbonate make up ca. 80 per cent of the mineral by weight. Oxides of silicon, aluminium and iron comprise the bulk of the remaining 20 per cent, with other components in small amounts.

| | |
|---|---|
| Calcium carbonate | $CaCO_3$ |
| Magnesium carbonate | $MgCO_3$ |
| Silica | $SiO_2$ |
| Alumina | $Al_2O_3$ |
| Iron oxide | $Fe_2O_3$ |
| Sulphate | $SO_3$ |
| Phosphorus | $P_2O_5$ |
| Potash | $K_2O$ |
| Soda | $Na_2O$ |

When limestone is crushed and heated in a kiln to 1450°C the carbonates decompose, releasing carbon dioxide, $CO_2$, and forming oxides CaO and MgO.

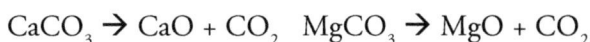

$$CaCO_3 \rightarrow CaO + CO_2 \quad MgCO_3 \rightarrow MgO + CO_2$$

While the carbonates are very stable, unreactive materials (hence forming minerals existing for hundreds of millions of years in the earth's crust) the oxides are quite reactive. CaO is what is generally known as lime, and when added to water forms the strongly alkaline calcium hydroxide, $Ca(OH)_2$.

Multiple species form in the kiln where the limestone is heated, as a result of reactions between CaO, alumina, silica and iron oxide. The four main chemical compounds that are present in the product are tricalcium silicate $(Ca_3SiO_5)$, dicalcium silicate $(Ca_2SiO_4)$, tricalcium aluminate $(Ca_3Al_2O_6)$[10] and tetracalcium aluminoferrite $(Ca_4Al_2Fe_2O_{10})$. The amounts of these species can be varied by addition of clays rich in desired elements. The reactive mixture from the cement kiln is known as clinker. Varying the proportions of these compounds affects the properties of the resulting concrete. All of the reactions lead to the production of calcium hydroxide and heat, with typical reactions being

$$2Ca_3SiO_5 + 7H_2O \rightarrow 3CaO.2SiO_2.4H_2O + 3Ca(OH)_2 + Heat$$

$$2Ca_2SiO_4 + 5H_2O \rightarrow 3CaO.2SiO_2.4H_2O + Ca(OH)_2 + Heat$$

The different compounds present in the clinker react with water at very different rates, and hence make distinct contributions to the properties of the resulting concrete. The rapid reaction of tricalcium silicate with water produces a lot of heat (exothermic reaction) and forms calcium silicate hydrate which has high strength and contributes most to the early strength of cement hydrates. Dicalcium silicate reacts much more slowly with water so that any heat developed is dissipated before excessive temperature rise occurs. Similarly to tricalcium silicate, tricalcium aluminate reacts rapidly with water in a highly exothermic reaction. The consequences of these differences in reaction rates are that the properties of the resulting concrete can be tuned by varying composition in the kiln, and that the strength of concrete continues to increase over time, considerably in the first four weeks and gradually over the following year after it is cast.

After the water is added to the cement and aggregate, it is essential that the mixture remains liquid for long enough so that it can be poured into moulds to set into the desired shape. The most rapid reaction is that of tricalcium aluminate, and this reaction needs to be retarded to maintain the ability to work with the liquid concrete. It was found empirically that this is accomplished by addition of gypsum, calcium sulphate, another mineral with widespread availability, of formula $CaSO_4$. It is now understood that this reacts with some of the tricalcium aluminate to form tiny crystals of ettringite, $Ca_6Al_2(SO_4)_3(OH)_{12}$, which cling to the surface and slow down the reaction in the early stages.

Various materials are added to the clinker to form concrete. The Romans used pozzolans, ash from volcanic eruptions, most cement uses sand, and a variety of other materials are in use as well. Most of this is filler and does not contribute to the strength of the concrete, but a portion of it is crucial. Sand is primarily silica, and the calcium hydroxide formed from the reaction of the clinker species binds with the sand in a reaction that is

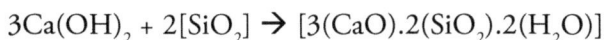

$$3Ca(OH)_2 + 2[SiO_2] \rightarrow [3(CaO).2(SiO_2).2(H_2O)]$$

This is sometimes known as the pozzolanic reaction. It is this material, coating the sand and linking the network of sand particles together, that is what is known as concrete, with its high compressive strength. For this to work, however, the size and shape of the sand particles is important. Experimentally it is found that desert sand, the most abundant, is generally too fine and smooth to be suitable for concrete formation. One way of understanding this at a macro level is that small, round, smooth particles are more easily rearranged under stress than larger irregularly shaped sand. At a slightly more molecular level, rounded particles would lead to a network that is similar in all directions, whereas sand particles with sharp angles would lead to bonding in many different directions, creating a stronger network. The ironic result is that countries such as Saudi Arabia and the UAE are importers of sand for use in construction.

What is clear from all of this chemistry is that production of cement from minerals produces carbon dioxide as a product and also requires a lot of heat, historically always generated by burning fossil fuels. It is estimated that 4–8 per cent of global greenhouse gas emissions come from concrete production. As with several materials, the amount used increases as an economy develops, eventually reaching a peak after which economic development is decoupled from materials use. This is clearly demonstrated for concrete and is also true for many metals.[11] Nonetheless, China produced and consumed more concrete in 2017–18 than the US used in the entire 20th century. In Chapter 9 ideas about the future of concrete and its environmental impact are discussed.

## TECH TALK 8 STEEL

Steel is an alloy of iron with small amounts of carbon, generally between 0.1 and 2 per cent. There are many varieties and grades of steel, with distinctive properties, made by adding other metals in small amounts.

## PROCESSES

There are many books dealing with the processes for making steel, their history and relative merits. Hence there are only a few brief notes about this here.

What happens in the most dominant steelmaking processes is that molten iron ore is heated in the presence of an oxidising agent to drive off impurities, leaving a controlled amount of carbon. It is the carbon that determines the hardness and other desirable properties of basic steel compared to iron.

Although Henry Bessemer, when he invented the blast furnace for making steel in the 1850s, realised that it would be better to use pure oxygen rather than air, at that time, and for nearly a century thereafter, large quantities of pure oxygen were not available commercially to support steelmaking. By 1940 this had changed, a result of the work of Linde, Hampson and others during the early 20th century on cooling air until it liquefied and separating the oxygen from the nitrogen. Once the basic oxygen furnace for steelmaking was commercialised it drove all other methods out of the market, at least for a time.

In all the air/oxygen-based approaches for making steel, the key was recognition that the reactions driving off impurities as gases are themselves exothermic, that is, they produce heat. To be economic, processes must effectively capture that heat and use it to keep the iron molten.

After World War II increasing attention was focused on using electric arcs rather than combusted fossil fuels to produce the molten metal, still blowing in oxygen once the melt had been achieved. This turns out to be a particularly useful approach for smaller mills. It started with a desire to build up steel production capacity in war-ravaged European cities for reconstruction, but really took hold as recycling of scrap steel became an important component of the market. Today electric arc furnaces do most of the processing of scrap and some iron ore, and basic oxygen furnaces do most of the rest.

One small sector of the steel production is where iron is produced in a very pure, molten state as a by-product of other processes. As an example, the Quebec-based Rio Tinto subsidiary Fer et Titane (Iron and Titanium) processes an ore called ilmenite, of formula $FeTiO_3$, mainly to make titanium dioxide, $TiO_2$, for pigments and cosmetics. This is a special case of a small sector of the steel production known as direct reduced iron, the most basic example being converting iron oxide, $Fe_2O_3$, into iron. All these direct reduced iron processes for making steel use less energy than a conventional basic oxygen furnace, but when it is done as a by-product that is more valuable the process becomes quite attractive economically, though it will never achieve the scale because of the smaller quantities of ores like ilmenite compared to other iron ores. Still, as will be seen below, a process making iron from ilmenite produces a very high-purity iron (indeed it has been used as a food additive in the past) and so can be alloyed with other elements very precisely to produce the highest value steels.

## ALLOYS – THERE ARE MANY STEELS

Steel has so many different applications – from structural support of buildings and other infrastructure, containers such as paint cans, many automotive parts, the outer frame of white goods, parts of engines, knives and swords, kitchen cutlery and other utensils, wires and cables. Different properties will be of greater or lesser importance depending on the application. The most common properties that steel manufacturers try to adjust include strength, hardness, toughness (on impact), wear resistance, ductility, corrosion resistance, brittleness, hardenability, thermal and electrical conductivity.

Most of these need little explanation, though each has well-defined methods for measurement and scales for quantitative description.

Ductility is the ability of the steel to have its shape changed, for example to be pulled into a wire. Clearly, for structural applications ductility would usually be an undesirable property. Hardenability is different from hardness. Generally, as steel is produced in the molten state it is set into a form that is produced continuously, such as a circular or square cross-section, and then quenched in oil, air or water. When it is quenched, various reactions occur in the material, usually from the outside towards the inside, which cause it to become harder. In general, the hardness decreases from outside going in, and the ability to harden to greater depths is the hardenability. Each combination of properties makes a steel alloy more or less suitable for a particular application.[12]

Over millennia people did trial and error experiments with steel to manipulate these properties, so as to make sharper and longer-lasting swords, for example. As progressively more metals were made in pure form, and techniques for accurately blending these into steel were developed, the properties of alloys were developed. In the 20th century various techniques such as electron microscopy and X-ray diffraction allowed characterisation of the steels at the microscopic or even atomic level, and an understanding of what each additive was doing.

A large portion of the metallic section of the periodic table has been tried in terms of alloys, and many are employed today. Common metals used in steel include manganese (the most common one), nickel, chromium, molybdenum, copper, vanadium, silicon and boron. Less common ones include aluminium, cobalt, cerium, niobium, titanium, tungsten, tin, zinc, lead and zirconium.[13] With various combinations and quantities there are more than 3,500 different steel products in the market today.

The steel property most people encounter is corrosion resistance, universally known as stainless steel. The key component of stainless steel is chromium, generally present between 10 and 15 per cent. Stainless steels may also have added nickel, manganese and many other elements, including non-metals such as nitrogen and phosphorous. Each has its own effect on properties, both beneficial and detrimental, so they are balanced using a combination of science and empirical results. This table shows just how complex this is:

| PROPERTY | C | CR | NI | S | MN | SI | P | CU | MO | SE | TI/NB |
|---|---|---|---|---|---|---|---|---|---|---|---|
| Corrosion Resistance | – | ✓ | ✓ | ✗ | – | – | ✓ | – | ✓ | – | – |
| Mechanical Properties | ✓ | ✓ | – | – | ✓ | ✓ | ✓ | ✓ | ✓ | – | ✓ |
| High Temperature Resistance | – | ✓ | ✓ | ✗ | – | – | – | – | ✓ | – | ✓ |
| Machinability | ✗ | ✗ | – | ✓ | – | – | ✓ | – | – | ✓ | – |
| Weldability | ✗ | ✗ | – | ✗ | ✓ | – | ✗ | – | ✓ | – | ✓ |
| Cold Workability | ✗ | ✗ | ✓ | ✗ | – | – | – | ✓ | – | – | – |

✓ = Beneficial  ✗ = Detrimental

Effect of alloying elements on properties of stainless steel.[14]

In contrast to the large quantities of chromium added to deliver corrosion resistance, very small quantities (.001–003 per cent) of elements such as boron have a marked effect on hardenability. Likewise, copper in small amounts can improve corrosion resistance.[15]

## TECH TALK 9 FLOAT GLASS

When liquids are cooled to below their melting point, they usually form crystalline solids. This process is speeded up by addition of seed crystals, or small quantities of dust around which the crystals nucleate and grow. While in most cases the liquids are disordered, the crystalline solids generally have ordered three-dimensional structures, the exception on the liquid side being the class of materials known as liquid crystals, which are ordered fluids.

For most liquids it is possible to cool them in such a way as to quench the disordered, amorphous state, though often this is metastable and will crystallise gradually over time. Polymeric materials are particularly interesting in this regard. For example, when bread is baked the starch-water gels that form from flour and water are quenched into an amorphous form when removed from the oven, this giving the texture of fresh bread, while stale bread is the more stable crystalline form that grows over time.

Amorphous solids, lacking three-dimensional order, are known as glasses and have desirable optical properties of high transparency to visible light. A highly ordered crystal, such as often found in large perfect gemstones, may also have such properties, but most crystalline solids are full of small crystalline regions joined together by defects which reflect, refract and scatter light. Glasses do have some of the flow properties of liquids, but, depending on the temperature, this may be very slow.

The best-known glass, which is stable to crystallisation over exceptionally long periods of time, is that primarily composed of the most common material, silica, common sand, and is of course called glass. If the silica had been allowed to crystallise it would be in a form such as quartz. The formulation of minerals to prepare it has also been known for thousands of years.[16] Soda ash (sodium carbonate) is added to sand to lower the melting point. Limestone (calcium carbonate) makes the glass harder and more durable, and dolomite (a mixed carbonate of calcium and magnesium) improves weathering properties and also makes the glass more workable in its molten form. Various other additives and coatings are used for specific applications. One major specialty glass composition created early in the 20th century in both Germany and the US was borosilicate glass, in which boron oxide was incorporated. This was the basis for glass that resisted thermal shock, marketed in the US by Corning as Pyrex and in Germany by Schott as Duran. Besides its role in cooking, it revolutionised laboratory chemistry which often involves heating and cooling of reaction mixtures. As with steel and cement, glassmaking is an energy intensive process, which, as in the case of cement, also involves release of carbon dioxide as part of the chemistry.

As the market for glass in buildings and houses expanded, the process for making it by casting large sheets was a bottleneck. From a process point of view, the problem of glass manufacture was thus very similar to that for paper, needing innovation so that a continuous ribbon could be made and cut rather than producing individual sheets. Glass also involved steps involving polishing to achieve desired optical properties. Similarly, continuous rolling of steel was important for certain uses. Henry Bessemer, whose inventions revolutionised steel production, also developed methods that were commercially successful for rolling steel. Bessemer, who was a prolific inventor, patented a method for making continuous ribbons of glass using molten metal flotation in 1848, but it was not commercially successful, and he abandoned it.

This idea was taken up by Alistair Pilkington[17] (1920–1995) at UK glass company Pilkington Brothers in the early 1950s. In the process, the molten mixture is first prepared in a furnace at 2000°C, then, as it becomes more homogeneous and bubbles are eliminated, the temperature is reduced to 1100°. It moves out of a ceramic spout very gently onto the surface of a bath of molten tin. The ribbon is held at a high enough temperature over a long enough time for the irregularities to melt and for the surfaces to become flat and parallel; because the surface of the molten tin is flat, the glass also becomes flat. The ribbon is then cooled down to 600° while still on the molten tin, until the surfaces are hard enough for it to be taken out of the bath without rollers marking the bottom surface. In this way a glass of uniform thickness and with bright, fire-polished surfaces is produced without the need for grinding and polishing. It took about seven years until the first demonstration scale plant could be built, and then 14 months until it worked well enough to produce defect-free glass.

Initially the process produced glass 6 mm thick (which met the needs of about half the market at that time) but it was found that by controlling the speed at which the (up to 3 metre-wide) ribbon is withdrawn from the bath that the thickness could be varied between 0.4 mm and 25 mm. This variation in thickness of the product is possible without compromising the quality of the surface. If process conditions are tightly controlled glass that is completely free of imperfections can thus be produced without any post polishing steps, and in very high yield. Coatings can be applied to the glass using chemical vapour deposition, for such properties as anti-reflection, ultraviolet light absorption and tinting. After removal from the bath, the glass still may develop strains as it cools. These are routinely removed by heat treatment in post-production ovens, a process step known as annealing. Today, a float glass plant can operate non-stop, producing 6,000 km of glass per year, for more than 10 years without needing any shutdown.[18]

## TECH TALK 10 ALUMINIUM BY VICTORIA STOMBERG

*My first thought was I had laid my hands on this intermediate metal which would find its place in man's uses and needs when we would find the way of taking it out of the chemists' laboratory and putting it in the industry.*

Preface of *Aluminium, its properties, manufacture and applications,* book written by French chemist Henri Étienne Sainte-Claire Deville in 1859

Aluminium is the most abundant metal and the third most abundant element of all elements in the earth's crust. It is very light, with a density around one-third that of steel or copper. It is corrosion resistant (because it is immediately coated on the surface with a thin layer of aluminium oxide which adheres, protecting the inner material) and is an excellent conductor of electricity and heat. These properties make it very attractive for numerous applications. In the Earth's crust, aluminium does not exist in its metallic form, but is only found in combination with other elements in ores. Bauxite is the primary ore of aluminium which contains many different metallic compounds but consists of 45–60 per cent aluminium oxide (alumina).

## FROM DISCOVERY TO COMMERCIAL PRODUCTION

Aluminium oxide is an ionic compound and very stable. Each aluminium ion is bonded to six oxygens. Although aluminium salts known as alum have been used since antiquity, it took 80 years from the proof of the metal as a distinct element by Humphrey Davy in 1807 until a process was developed to efficiently extract metallic aluminium from alumina in 1886. In 1845 German chemist Friedrich Wöhler established many of the attractive properties which encouraged other scientists to find a way to extract aluminium from alumina. Then in 1854 Henri Sainte-Claire Deville developed the first commercial process to produce the metal, but the reduction process using aluminium chloride and sodium metal was very costly. Nonetheless this process was copied throughout Europe and allowed scientists to produce aluminium on a kilogram scale.

The process that is still used today to extract aluminium was discovered in 1886 by two scientists almost simultaneously but independently in the United States and France. Only a slight advance was made by Karl Bayer (founder of the Bayer Company) who invented an improved process for making aluminium oxide from bauxite. The combination of these two developments reduced the cost of aluminium to approximately 20 per cent that of Deville's price, making the metal a commercial commodity.

## TWO SCIENTISTS, ONE PROCESS

Charles Martin Hall (1863–1914) developed the smelting process in Ohio while Paul Héroult (1863–1914) invented the same process in Normandy. In this process aluminium is extracted from alumina by electrolysis. For the electricity to move through the cell alumina needs to be molten and so is dissolved in molten cryolite ($Na_3AlF_6$) to lower the melting point (melting point of alumina: >2000°C vs. melting point when dissolved in cryolite: 1000°C). Up to 25 per cent of alumina is dissolved which is enough to decrease the melting point. Using cryolite as a solvent reduces the energy costs involved in the extraction of the metal. An electric current is passed through the mixture and the ions in aluminium oxide move freely inside the cell.

## THE SMELTING PROCESS

The diagram overleaf shows an aluminium oxide electrolysis cell. In the electrolysis process for producing aluminium both electrodes (positive electrode or anode and negative or cathode) are made of graphite. The anodes are lowered into the cell vertically. The cathode is the lining of the cell. The dissolved aluminium ions ($Al^{3+}$) collect at the cathode where they receive electrons and are reduced to atomic aluminium. The molten aluminium then sinks to the bottom as its density is higher than that of molten cryolite at the required temperatures and is tapped off. Alumina is added to the cell as the aluminium is removed. The oxide ions in alumina collect at the anode where they lose electrons and are oxidised to oxygen gas which then reacts with the carbon anode, forming carbon dioxide. This means the anodes are consumed during the electrolysis process and need to be replaced regularly. The cathode also degrades during electrolysis but much more slowly and needs to be replaced roughly every two to six years.

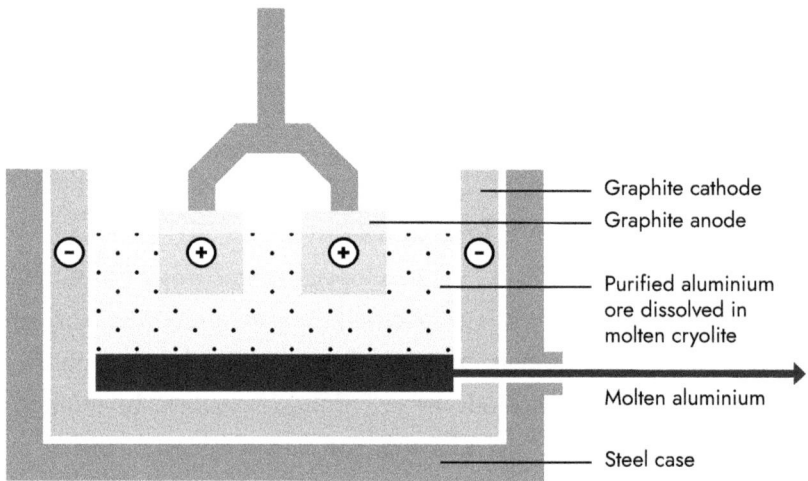

Graphite cathode
Graphite anode

Purified aluminium
ore dissolved in
molten cryolite

Molten aluminium

Steel case

## THE CHALLENGE

To obtain electricity in the small American college town Oberlin in the 1880s Hall had to build and assemble batteries. Hall and his adviser on the faculty of Oberlin College, Frank Jewett, used Bunsen-Grove cells, which consist of a large zinc metal electrode in a sulfuric acid solution that surrounds a porous ceramic cup containing a carbon rod immersed in concentrated nitric acid. The amount of electricity needed for the electrolysis process of aluminium required a lot of these cells to be assembled, which was a great challenge at that time. The supplied electricity was still limited, and external heating was required to keep the reaction mixture molten, making the process very energy intensive. The eventual successful laboratory experiment took place on 23 February 1886. Jewett confirmed that what Hall had produced was aluminium and Hall applied for a patent on 9 July 1886. His success was not only a result of Hall's stamina and inventiveness but also of the support he received from Jewett, who supplied materials for the Bunsen-Grove cells. Additionally, companies in Cleveland supported Hall's work by supplying him with gasoline for the furnace from Standard Oil, sulfuric acid and concentrated nitric acid for the batteries from Grasselli Chemical, and graphite rods for electrodes from Brush Electric.

At the same time Paul L. T. Héroult was granted a French patent in April 1886 for a very similar process using alumina and cryolite. In May 1886 Héroult applied for a US patent. Hall, who had made aluminium using this method before the date of the French patent, had to prove that his work had precedence to obtain patent protection in the United States. He was able to do so because his family and Jewett could confirm his success. Postmarked letters that he had sent to his brother served as evidence and helped him get the patent rights for the electrolysis process of aluminium in the United States.

## INVESTMENTS AND FURTHER IMPROVEMENTS OF THE PROCESS

Hall did not have the financial resources to develop his process further and commercialise it. With the help of his brother, Hall got financial support from Henry Baldwin and Simeon Brown so he was able to develop his process in Allston, Massachusetts, where he had a small dynamo to supply the needed electricity. Hall had difficulties developing the process further and upscaling it, so the brothers Baldwin and Brown withdrew their support. He worked on the process with his original equipment in Oberlin until he managed to get support from the Cowles brothers, who had a metal business called the Cowles Electric Smelting and Aluminium Company in Cleveland and in Lockport, New York. Hall was sent to Lockport to work on his process, but the Cowles brothers were disappointed with his progress and terminated the agreement in 1888. Fortunately, Romaine Cole and Alfred Hunt saw potential in Hall's process and by August 1888 Hunt had secured funding for a pilot plant on Smallman Street in Pittsburgh, Pennsylvania, which was fitted with a large, steam-engine-powered

Westinghouse dynamo to provide the electricity for the electrolysis process on such a large scale. With a scaled-up process using external heating, Hall and his new assistant, Arthur Davis, were making aluminium by November 1888. Scaling up seemed to not be a problem anymore but soon Hall realised that external heating meant the reaction pots had short lifetimes, so he switched to internal heating. Another improvement was the use of open steel pots lined with graphite which were much more durable and made it easier to adjust the anodes that were consumed during the process. These developments were great milestones for a process that is still used today.

## EXPANSION OF ALUMINIUM PRODUCTION

The small pilot plant on Smallman Street was closed in 1892 as the markets for inexpensive aluminium grew, and a new site in New Kensington north of Pittsburgh was opened. Not long after another plant was opened at Niagara Falls, New York, that was using hydropower to generate the needed electricity as a more economical alternative to steam-powered electricity generation. Other plants for aluminium production were opened in Quebec in 1899 and in western New York 1903 that all used hydropower.

## COMMERCIALISATION AND APPLICATIONS OF ALUMINIUM

Before 1886 there were on the one hand precious metals like gold, silver and platinum that were too expensive for everyday uses. On the other hand, there were less expensive metals like iron and copper which had multiple uses but were subject to corrosion. The establishment of inexpensive aluminium production and its attractiveness led to a variety of applications where we still use aluminium today.

Once aluminium production was established with cheaper electric power the metal became an inexpensive commodity. While it was hard initially for aluminium manufacturing businesses to prosper because markets for aluminium did not exist, only a few of the competitor businesses where aluminium could displace them survived into the 20th century. Aluminium was used to replace copper, brass and bronze. It is a good electrical conductor and extremely lightweight.

The corrosion resistance makes it a suitable material for a variety of products such as cans, kitchen utensils, window frames, aeroplane parts and engine blocks. Aluminium itself is not particularly strong, but alloys of aluminium with copper, manganese, magnesium, zinc and silicon are still lightweight but stronger. These alloys are used in many forms of transport, where the light weight means reduced fuel consumption. Aluminium has been used in aviation even before aeroplanes had been invented. Count Ferdinand Zeppelin used aluminium as a construction material for his famous Zeppelin airships in the 1890s, which was one of the first large-scale applications of aluminium in Germany. In early aeroplanes aluminium was only used for some parts to reduce the weight but gradually replaced the cloth, wood, steel and other parts in the early 1900s. The first all-aluminium plane was built in the early 1920s. Even inside aeroplanes aluminium is used to further save weight, reduce fuel consumption and increase the load capacity

One of the most important uses of this metal is long-distance electric power transmission. In the early years of the 20th century power lines were developed that were made from an aluminium alloy wrapped around a steel interior to provide added strength. These still make up our electric grid today. The great strength-to-weight ratio that aluminium possesses makes it suitable for applications where high strength and low weight are required.

Aluminium is widely used in food and pharmaceutical packaging as it is non-toxic and does not leach or taint the products with which it is in contact. Preserving drinks using aluminium cans was one of the greatest innovations in food packaging in the 20th century. The first aluminium cans were introduced to the market in 1957 as a better alternative to steel (tinplate) cans. In addition to its other properties, aluminium cans have a clean appearance and are more malleable, making it more attractive than steel. Another application where aluminium has replaced tin is aluminium foil, which is still often referred to as 'tin foil'. The first aluminium foil was made in 1910 and it is used as food packaging, insulation, electromagnetic shielding and in cooking. Again, aluminium foil is more malleable than tin foil and tin foil tends to give a slight tin taste to food wrapped in it which aluminium foil does not. These properties do not degrade when aluminium metal is recycled and the recycling process only requires around 5 per cent of the energy input required to produce new aluminium metal, making recycling of this material very attractive and sustainable. The benefits that can be obtained from recycling are influenced by the purity level of the aluminium scrap. Hence cans where different alloys are used for the top versus the sides are less easily recycled. Used in so many everyday applications aluminium has become a material we cannot imagine living without.

## TECH TALK 11 ELECTRICAL INSULATORS – BAKELITE AND THE BEGINNING OF PLASTICS

Having made a substantial amount of money from the sale of his photographic paper product to Eastman, Leo Baekeland started to look at other interesting areas of chemistry. He first pursued electrochemistry, and made developments in several processes that led to the formation of Hooker Chemical Company, which used the newly available electricity at Niagara Falls to power its process for converting salt (sodium chloride) into chlorine and sodium hydroxide (caustic soda or lye). With his wealth, Baekeland built a private laboratory just north of New York City and began to explore new areas for research. He was always driven to find areas of chemistry from which he could make money.

His target was a synthetic material to replace shellac, which was finding use not just in the rapidly growing field of electrical insulation but as a material for phonograph records. Baekeland was aware that the great German chemist Adolf von Baeyer and his students, in their search for synthetic dyes, had mixed phenol (also called carbolic acid) and formaldehyde as well as other aldehydes but obtained a black goop that they considered to be of no use. In his private laboratory, he took this as his starting point and made a systematic investigation of the reaction and its products, varying temperature, pressure and various additives. By 1907 he had succeeded in making the first plastic that could be moulded into a fixed shape at reasonable temperatures, ca. 150°C, and hence called a thermoset plastic.

It had superb properties of electrical insulation. From the prior experience of his inventions Baekeland was very skilled at protecting his intellectual property and immediately filed a patent application, which was granted in 1909. It was the first of more than 100 patents he obtained in the area of phenol formaldehyde resins. He named the product after himself, Bakelite, and in 1910 set up a company, General Bakelite,[19] to scale up production.

Bakelite found its way into many applications, from the distinctive black case of the rotary telephone to cigarette holders, but its excellent electrical insulation properties meant it could replace shellac as an insulator and be used in many other new electrical devices, such as radios. All of the early uses of Bakelite were in electrical insulation, including bases for light bulbs, vacuum tubes, bushings in motors and automobile distributor caps. Moreover, his success inspired a wave of organic synthetic efforts to produce novel polymeric materials for applications across diverse industries.

## TECH TALK 12 CHEMISTRY OF NATURAL AND SYNTHETIC RUBBER[20]

The rubber tree, *hevea*, native to Brazil, produces a white watery liquid when its bark is cut, which contains a polymer of the molecule isoprene, a five-carbon molecule with two double bonds.

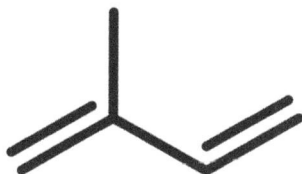

The most basic reaction that occurs is between a natural product of biological processes in the *hevea* tree producing a phosphate that contains isoprene and a pure isoprene molecule. This starts a reaction leading to a long chain polyisoprene. The double bond provides a rigidity to the backbone, with the methyl, $CH_3$ groups, all on the same side, while the single bonded section of $C-H_2C-CH_2-CH$ provides flexibility and rotation out of the plane. It is a characteristic of biological processes of all sorts that they produce long-chain molecules with very particular structural features, such as right- and left-handedness, helicity, folding. In this case the stereo-regularity is that the methyl groups are all on the same side – other possibilities being that they alternated up and down, or that their position was random. In fact, isoprene can be polymerised to give four different chemical structures,[21] of which only one is produced by the rubber tree.

Some of these biopolymers, like polyisoprene, appear as a milky liquid that humans learned to harvest without killing the tree, and proved to be of great utility. What is unusual about the polyisoprene polymer produced by *hevea* is that there are laboratory invented processes that can produce the identical polymer very readily and in commercial quantities, although this was only accomplished in the 1950s using a catalyst invented by the German scientist Karl Ziegler and first used to produce stereo-regular polymers by the Italian Giulio Natta. For most biological processes it has proved very difficult to achieve the same selectivity in the laboratory that biology has evolved, and this was the case for many decades with polyisoprene.

The liquid from the tree, known as latex, had uses and was known outside of Brazil in the 1700s, mostly as novelty items such as bouncing balls. Indeed, it was used by natives in Brazil in ball games. The first practical use was proposed by the great English scientist Joseph Priestley in 1770, who wrote, 'I have seen a substance excellently adapted to the purpose of wiping from paper the mark of black lead pencil.' He called the substance rubber. In the same year this began to be produced and sold commercially for this purpose in England, the first practical use of rubber in Europe.

The major advance in new uses for rubber was vulcanisation, involving the addition of sulfur. Chemically, this is a complex set of reactions, especially in the early days when it was more cookery than science. What is generally agreed is that vulcanisation chemistry involves sulfur forming bonds between the polymer chains.

Vulcanisation causes rubber to shrink while still retaining its original shape. The vulcanisation process also hardens the rubber, making it less susceptible to deformation – particularly compared to non-vulcanised rubber which will deform far more quickly under stress. This hardening of the rubber also increases its tensile strength. Further benefits of vulcanised rubber are that it returns to its original shape when it is deformed, has low water absorption, high resistance to oxidation and abrasion, and resistance to organic solvents. All of these properties are important for automobile and truck tyres, though these were not required for another 60 years after vulcanisation was invented.

While synthetic isoprene was not produced until after World War II, there was a great deal of effort by chemists to produce other synthetic rubber polymers during the 1920s and 30s, and this intensified as the possibility of war became greater. By this time, most of the rubber, particularly for Europe, was coming from Southeast Asia rather than Amazonia. It became clear that the Japanese would be able to cut off supply. The story of synthetic rubber thus parallels exactly that of the need for synthetic substitutes for silk.

One possibility was to polymerise isobutylene:

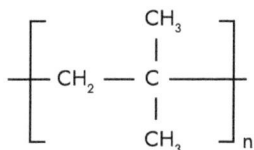

$$\left[ \begin{array}{c} CH_3 \\ | \\ CH_2 - C - \\ | \\ CH_3 \end{array} \right]_n$$

This makes what is known as butyl rubber, and while it lacks the hardness of vulcanised rubber it was found to have very low permeability to air, making it an ideal material for inner tubes of tyres. Before its use became widespread it was necessary to top up the air pressure in tyres at least monthly.

In the laboratories of IG Farben and Bayer, in 1930s Germany, Herman Mark and colleagues developed a new rubber polymer of completely different chemistry, introducing nitrogen in the form of nitrile groups, $C \equiv N$.

This rubber was known as BunaN, and commonly called nitrile rubber. Because two different monomers are polymerised together to make the polymer, properties can be varied by adjusting the ratio in which they are used. Nitrile rubbers are particularly resistant to oils, so are used for hoses where oil products are being pumped as well as industrial use rubber gloves.

One other class of synthetic rubber (from a vast literature and commercial products) is made by introducing chlorine into the polymer, for example by polymerising the molecules shown below whose correct chemical name is 2–chlorobutadiene.

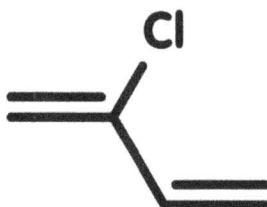

The generic name for such rubbers is polychloroprenes, and the DuPont product made from this monomer is called Neoprene. These rubbers are very useful for seals and gaskets.

The 20th century thus saw the opening of a vast area of chemical synthesis devoted to synthetic rubber materials. With practically each new polymer, it was possible to provide a better product for use in vehicles – the largest volumes being with tyres, but including seals, belts and hoses. In turn, each of these advances reduced the likelihood of a failure happening.

49 'Is Lab-Grown Leather The Future For The Fashion Industry?', *British Vogue*, 14 November 2020, https://www.vogue.co.uk/fashion/article/lab-grown-leather accessed 21 June 2021

50 See, for example, refs 3–5, and for even more speculative ideas J. F. Coates, J. B. Mahaffie and A. Hines, *2025*, Oakhill Press, 1997.

## Notes

1 BoPET, https://en.wikipedia.org/wiki/BoPET accessed 5 July 2021

2 Many of these advances are described in B. J. Bulkin, *Solving Chemistry*, Whitefox, London, 2019.

3 Warp and weft, https://en.wikipedia.org/wiki/Warp_and_weft accessed 16 July 2021

4 'John Kay and the flying shuttle', Stories from Lancashire Museums, https://lancashiremuseumsstories.wordpress.com/2020/05/22/john-kay-and-the-flying-shuttle/ accessed 16 July 2021

5 https://en.wikipedia.org/wiki/Spinning_jenny accessed 15 July 2021

6 For a detailed description of all the chemical reactions involved in the Kraft process see Henriksson, Gunnar, Germgård, Ulf and Lindström, Mikael E.. 'A review on chemical mechanisms of kraft pulping', *Nordic Pulp & Paper Research Journal*, vol. 39, no. 3, 2024, pp297-311. https://doi.org/10.1515/npprj-2023-0015 accessed 22 June 2021

7 The detailed chemical reactions occurring in the recovery boiler are described in 'Chemical Reactions in Kraft Pulping', http://www.h2obykl.com/images/Reactions%20in%20Kraft%20Pulping.pdf accessed 22 June 2021

8 https://commons.wikimedia.org/w/index.php?curid=74783698 accessed 22 June 2021

9 For an explanation see D. L. Beveridge and B. J. Bulkin, *J. Chem. Ed*, 48, p587, 1971.

10 These formulae look complex but are more easily understood if we see something like $Ca_3Al_2O_6$ as being 3CaO and one $Al_2O_3$, although chemically each species behaves differently from the sum of its parts.

11 Andrew McAfee, *More from Less,* Simon and Schuster, 2019 p75ff; and Julian

Allwood & Jonathan Cullen, *Sustainable Materials,* UIT, 2015, p287 ff.

12 An excellent explanation of hardenability, what it is and how it is measured, can be found in 'Hardenability of Steel', In The Loupe, https://www. harveyperformance.com/in-the-loupe/hardenability-of-steel/ accessed 15 June 2021

13 Alloy steel, https://en.wikipedia.org/wiki/Alloy_steel accessed 17 June 2021

14 Stainless Steels Alloying Elements, https://www.azom.com/article. aspx?ArticleID=13089 accessed 17 June 2021

15 'Common Alloying Elements for Steel and their Effects', Bortec, https://bortec-group.com/glossary/alloying-elements/ accessed 17 June 2021

16 For an excellent discussion of ancient glass production, see Stephen L. Sass, *The Substance of Civilization,* Arcade, 1998, Chapter 6.

17 Although this was a family company, Alistair Pilkington was not a member of the family. However, when he was hired, because of his name, someone decided that he should be on the 'family track' for advancement in the company, and this gave him scope to influence the direction of research while still relatively young. It probably also helped in his project to develop what became known as the float glass process not being cancelled after it went on for several years without success. (Private communication from Alistair Pilkington to the author in 1990).

18 Pilkington is currently owned by NSG, Nippon Sheet Glass Corporation. Their website gives a good description of all the steps in the float glass process, https://www.pilkington.com/en/global/knowledge-base/glass-technology accessed 20 January 2021

19 Bakelite Corporation was sold to Union Carbide in 1939.

20 There is a vast literature on rubber chemistry. Useful introductory articles include 'Natural Rubber: Structure and Function', Halcyon, The world's leading rubber franchise, https://www.halcyonagri.com/en/natural-rubber-structure-and-function/#:~:text=Natural%20rubber%20is%20a%20 polymer,it%20is%20built%20is%20isoprene and 'How is rubber Vulcanised? The chemistry and history', Martins Rubber, https://www.martins-rubber. co.uk/blog/how-is-rubber-vulcanised-the-chemistry-and-history/ accessed 20 January 2021

21 By Roland.chem – Own work, CC0, https://commons.wikimedia.org/w/index. php?curid=54275727 accessed 21 January 2021

# INDEX

9 781916 797840